IISS

GW00367347

STRATEGIC SURVEY

1994-1995

Published by  Oxford University Press for

The International Institute for Strategic Studies
23 Tavistock Street
London WC2E 7NQ

STRATEGIC SURVEY 1994–1995

Published by Oxford University Press for
The International Institute for Strategic Studies
23 Tavistock Street, London WC2E 7NQ

DIRECTOR   EDITOR
Dr John Chipman   Sidney Bearman

This publication has been prepared by the Director of the Institute and his Staff, who accept full responsibility for its contents, which describe and analyse events up to 31 March 1995. These do not, and indeed cannot, represent a consensus of views among the world-wide membership of the Institute as a whole.

Managing Editor: Rachel Neaman
Production Supervisor: Denise Fouché
Editorial Assistant: Karen Smith

First published May 1995

ISBN   0 19 878 1369
ISSN   0459 7230

Strategic Survey (ISSN 0459 7230) is published annually by Oxford University Press.
The 1995 annual subscription rate is: UK £19.50; overseas $US30.00.
Payment is required with all orders and subscriptions. Prices include air-speeded delivery to Australia, Canada, India, Japan, New Zealand and the USA. Delivery elsewhere is by surface mail. Air-mail rates are available on request. Payment may be made by cheque or Eurocheque (payable to Oxford University Press), National Girobank (account 500 1056), credit card (Access, Mastercard, Visa, American Express, Diner's Club), direct debit (please send for details) or UNESCO coupons. Bankers: Barclays Bank plc. PO Box 333, Oxford, UK, code 20-65-18, account 00715654. Claims for non-receipt must be made within four months of dispatch/order (whichever is later). Please send subscription order to the Journals Subscription Department, Oxford University Press, Walton Street, Oxford, OX2 6DP, UK, Tel: +44 (0) 1865 267907. Fax: +44 (0) 1865 267773.

US POSTMASTER: Send address corrections to Strategic Survey, c/o Virgin Mailing and Distribution, Cargo Atlantic, 10 Camptown Road, Irvington, NJ7 07111-1105, USA.

PRINTED IN THE UK by Bell & Bain Ltd, Glasgow.

CONTENTS

Perspectives

A pervading sense of impotence characterised international affairs during the past year. As a consequence it was a period of drift. If it could have been considered a period devoted to reflection as the world sought to adjust to the unsettled conditions with which it was struggling, it might have been a time well spent. Unhappily, however, it was simply one of inaction. The world appeared to be marking time, while many of the positive advances that had been made since the end of the bipolar geopolitics that had dominated the past 45 years receded. All too often the question was: who's in charge? And no answer came back.

Nation-states, as the world has known them for centuries, are facing a crisis of governance. In most countries on all five inhabited continents weak leaders are balanced uncomfortably atop shaky governments. Busy looking over their shoulders at disgruntled electorates, or under attack from rebellions of the left and right, they have had little time or inclination for strong or positive initiatives in foreign affairs. Even when dealing with domestic issues they have been reluctant to march out in front and have, at best, been content merely to sustain an unsatisfactory status quo.

When leaders have been assertive, they have often miscalculated the correct balance between the use of power and reliance on diplomacy to serve foreign-policy outcomes. Russia's invasion of Chechnya was one such case, but in Algeria's reaction to its Islamic problems, the West's policies towards the conflict in the former Yugoslavia, and US internal deliberations about how to solve the problem posed by Haiti, the world observed the difficulties leaders face today in balancing the threat or use of force with diplomacy and negotiations to achieve desirable policy outcomes.

Since so much foreign policy today is related not only to individual domestic policy, but also to the domestic policy of target states, the development of strategy has become complicated by the perceived need both to influence the leaders of other states and to address the populations they govern, with or without their consent. Thus today, in a world growing more pluralist, foreign policy requires what in the past it sought to avoid: entanglement in calculations about the internal affairs of other states. In assessing the effects of the use of force, or the utility of either sanctions or economic aid, leaders must determine the domestic positions of others. This so complicates strategic action that leadership and control of its consequences are notably absent.

International organisations were naturally infected by this malady of weak leadership. This was partly because those upon whom they depend for money, resources and troops had turned their eyes inward to better satisfy their own constituencies. Partly it was a result of the petty quarrels among these same weak leaders that prevented vacancies from being filled for far too long, and then filled not by a first-choice leader, but by someone who inspired the least negative reactions.

If this was true of international organisations, it was also true of alliances. Many alliances in 1994 were more involved in consensus-building, which normally meant inaction while it was under way, than in pursuing strategic goals. In 1994 the trend begun in the previous year was continued: since analysis of the causes of security problems now often differ, addressing security challenges can threaten the cohesion of alliance structures. The North Atlantic Treaty Organisation (NATO) was almost torn apart by differences of opinion over Bosnia, while the Association of South-east Asian Nations (ASEAN) avoided confrontation only by not commenting on China's attempt to advance its claims in the Spratly Islands at the expense of an ASEAN member.

States like China, and in a different way North Korea, sensing that a fear of the use of force provides unexpected freedom of action to those who would challenge timidity, felt their bargaining power rise. Only in the Persian Gulf, where at least the Anglo-Saxon powers were united, did tough talk marry up with tough action: the US and the UK continue to deter Iraq and contain Iran. At least here a long-standing security interest was defended with old-fashioned methods. But outside the Persian Gulf there were few instances where leadership, either for a defensive or positive purpose, was so regularly demonstrated.

The Dimensions of the Problem

There undoubtedly have been many historical periods in which this phenomenon has occurred, but in the modern era, since the end of the Second World War, drift has never been so evident. For better or for worse, the United States, whether under a Democratic or Republican president, was in the past prepared to step forward and take charge. The decisions taken may not always have been unanimously welcomed, not even by its allies. But decisions there were, and they led to action.

What has changed, of course, is that the reason for most of the action, or more accurately re-action, has disappeared. With the dissolution of the Soviet Union, the impetus for, and the consensus about, foreign-policy initiatives has also dissolved. Yet this can only be part of the answer. It has often been said that after the end of wars, or the destruction of a coherent world order, there is a four- or five-year period of transition while a new order is being constructed. After the Second World War there was a four-year hiatus before the need for new structures became clearer and leaders appeared to help in their construction. Now, at least five years after the end of the Cold War there is still no order and certainly no leadership. And in fact, the spread of weak leadership is a symptom of a deeper malady, reflecting more the changing nature of the nation-state than the dearth of talented individuals.

Any head of state who wants to strike out with new foreign-policy initiatives must buck and invigorate a reluctant following. For example, while polls show that more than half of the US electorate still believes that the United States ought to, and will, play a leading role in world affairs, these same polls indicate that foreign affairs barely register on the list of what the voters believe should be priorities. With a sigh of relief Americans feel that the end of the Cold War means that they need not continue lifting most of the international burden. They want their government to turn its attention, and money, to solving domestic problems, of which there are all too many.

President Bill Clinton is very much in tune with these sentiments. He ran for office on a platform that reflected these concerns. He is himself more interested in domestic than foreign affairs, and believes that it is time for a shift of emphasis and focus. Even if he thought differently, his political antennae quiver in rhythm with the peoples' moods. Unlike one contender for leadership who coined the aphorism that 'politics is the art of making the necessary possible', Clinton acts on the older, much more limited, dictum that at best politics is simply the art of the possible. This is compounded by a sense that for Clinton, the end of politics is often merely the achievement of a consensus rather than the implementation of policy. All too often his own policy goals were rendered unrecognisable by the process of negotiations that passage of the acts required. Now facing a resurgent Republican Party, and with an anxious eye on elections due in 1996, he is loath to spend the political capital necessary to create a consensus for leadership in foreign affairs that he does not believe already exists. The irony is that evidence suggests that this formula could win the prize he seeks.

In the latter part of 1994, President Clinton began to take a more direct role in foreign affairs. He lobbied hard for completion of the Uruguay Round of the General Agreement on Tariffs and Trade (GATT) and the formation of a World Trade Organisation (WTO), and succeeded. After almost a year of dithering he backed an invasion of Haiti and this more forceful response, however belated, was widely welcomed in the US. Although it again required the help of former President Jimmy Carter, the administration took a leading role in the difficult negotiations with North Korea which led to an agreement that might well halt its nuclear-weapon programme. And Clinton responded, some said over-responded, to an Iraqi movement of troops towards Kuwait by deploying a large contingent of troops by air. This decisive action resulted in an Iraqi retreat and, of course, was widely approved in the US.

The Republican Party sweep in the November mid-term elections brought Clinton's brief stab at leadership in foreign affairs to a close.

Although it seemed most likely that it was the people's unhappiness with the Democratic domestic agenda that fuelled the Republican surge, the President in the next six months did not attempt to revisit his foreign-policy successes with further initiatives.

At least the President continued to assert support for important international organisations, like the United Nations (UN), and for US efforts to provide aid to the less fortunate countries of the world. Insofar as they showed any interest in foreign affairs, the Republican leadership provided mostly a negative approach. While not exactly isolationist it signals a desire to rein in spending abroad, reduce support for the UN and cut grants, loans and other mechanisms which past administrations have championed as a way of supporting US aims in the world. That the administration would not put its loan guarantees to Mexico to the test of a vote it feared it would lose was a harbinger of the kind of inaction that could be expected from the ideologically driven leadership in Congress. The auguries for US leadership before the next election in 1996 are very poor indeed.

How About Europe?

Nor was there a leader, or government, in Europe that could fill the gap. Most of the major powers suffered the same disabilities as the US model, and many of them were in even worse shape. UK Prime Minister John Major's popularity with his people was so low that there was considerable doubt as to whether he would lead his party into the next elections. Scandal-ridden Italy was, in effect, without a government. During 1994 France had a government of cohabitation, with a seriously ill Socialist President and an uninspiring conservative Prime Minister. And in Germany, although the parliament gave Chancellor Helmut Kohl a fourth term, his margin of victory was paper-thin. The leadership of the European Union (EU) was weakened by internal squabbling over a replacement for Commission President Jacques Delors.

These governments, and the institutions of which they were members, focused most of their attention inward. Some attention needed to be paid to EU affairs as three countries (Austria, Finland and Sweden) voted to become new members, and to prepare for the Inter-Governmental Conference (IGC) which will take place in 1996. Adjustment of the EU's institutions will be a major question as newer nations carved out of the former Soviet Union clamour for entry. But it was only a slight exaggeration to say that the horizon of attention was not lifted much beyond Europe itself.

Even for those crises within Europe's physical boundaries there was neither much new thinking nor any effective new action. The war in Bosnia-Herzegovina ground on, threatening in the early part of 1995 to flare up again. With UN and EU peace plans alike stuffed into the dead-letter box, a new *ad hoc* organisation, the Contact Group, was established. It drew four of its five members from Europe (France, the UK, Germany and Russia) with the US as its fifth member. Its overt purpose was to rejuvenate the dormant peace process; its more hidden agenda was to try to unify the approaches of the outside powers.

In neither respect was it successful. Although it created a new peace plan, built on a geographic reconfiguration of Bosnia (the so-called Contact Group Map), this was no more welcome to the fighting forces than earlier outside attempts at mediation. Under considerable pressure from the US, the Bosnian government accepted the plan, but the Bosnian Serbs balked. In an effort to bring the Bosnian Serbs to heel, the Contact Group promised Serbian President Slobodan Milosevic relief from the sanctions which were grinding down the economy of the Federal Republic of Yugoslavia if he managed to get agreement from the Bosnian Serb leadership in Pale. Although Milosevic seems to have agreed to pressure the Bosnian Serbs, this ploy has not moved them. In mid-March 1995, the war, which had been merely smouldering during the course of a four-month cease-fire, blew up again.

Since the Contact Group plan failed even to convert the temporary cease-fire to a more permanent status, it is not surprising that it also failed to unify the views of the five members of the Group. They had for long been on different sides of the war. The US had always doubted the sincerity of the Serbs, believing that they had not only begun the war, but had no intention of allowing it to end except with their own victory. Thus Americans, and particularly Republican Members of Congress, were behind the notion that the embargo on arms to the Bosnian government should be lifted, and that air strikes against the Serbs should be reinstituted with the intention of 'levelling the playing field'. The West Europeans were not interested in placing blame, or increasing the dangers faced by their troops engaged in the UN's humanitarian work by increased fighting. The Russians tended to see the Serbs as the put-upon party, and complained of the bias against them that Russia perceived, particularly in the US position.

The Contact Group is only the latest in a long line of would-be mediators that have failed. Neither the UN nor any of the many European institutions that have tried to play a constructive role in the former Yugoslavia have had any success in finding a solution that would appeal to the bitterly opposed parties to the conflict.

Nor have international efforts had much success in other areas of the world. Although they have undoubtedly saved lives through their humanitarian efforts in Bosnia, and in Somalia before they were forced to withdraw, UN efforts have mostly served to expose the weakness of the positions they have adopted. In Rwanda in 1994, and with regard to the developing threat in Burundi in spring 1995, the UN was even unable to raise a force that might have stemmed the massacre. The accumulated actions, and inactions, by the international community have undercut respect for it in exactly those areas of the world where it is most important to maintain that respect. For without credibility it will not be able to play a useful role in stabilising other threatening situations. The time may have come for all outside powers and institutions to reconsider their attempt to provide a buffer in these intractable wars.

The Russian Enigma

All countries are having difficulty adjusting to the new geopolitical era, but Russia is having more difficulty than almost any other. Despite the weakness of its post-Soviet institutions, its military forces, its economy and its government, Moscow still insists on the status of a great power. President Boris Yeltsin, with the hot breath of the radical nationalists on his neck, not only demands that Moscow's position towards the Serbs throughout the former Yugoslavia be considered, but insists on a role with regard to NATO's plan for expansion to include former Warsaw Pact states.

The West thus faced the dilemma of how to enlarge NATO without involving Russia, and how to reassure Russia without giving it a veto over NATO policies towards Eastern Europe. By mid-March 1995, it appeared that the approach would be to develop a treaty arrangement with Russia that could serve as a *quid pro quo* for enlargement. But some were nervous about what the context of such a treaty could be, and uncomfortable with the notion that it was necessary for diplomatic purposes to pretend that Russia still had a great-power status that it barely enjoys in fact. Certainly, the idea of forging a 'strategic partnership' was difficult to advance at a time when Russia was engaged in disagreeable activities close to home.

The actions of the Russian government in the latter part of 1994 have seriously jeopardised the possibility of Russia evolving into a stable democratic nation. It badly mishandled economic affairs which led to a collapse of the rouble, and then it launched a disastrous military intervention in Chechnya. Civil control of the military forces seems to have been weakened. Efforts to reform the economy have been set back and President Yeltsin's own political position has been badly damaged.

These weaknesses are to no one's advantage. It is important for the West, as it is for the Central Europeans close to the Russian border, that Russia does not descend into chaos, or revert to authoritarian rule. Moscow's bumbling effort to bring Chechnya under its control has highlighted the extraordinary deterioration of what was once a highly professional, solidly trained and effectively armed force. Yet Russia still maintains a powerful, and possibly threatening, nuclear capability. And there are all too many in Russia who want to recover much of what was the Soviet empire.

For many years it has appeared that Boris Yeltsin provided the best bulwark against those who would act on such ideas. His behaviour throughout the past year, however, has thrown that judgement into doubt.

Those who want to help the Russian people to make the transition to a democratic, free-market society have a difficult balancing act to perform. They must continue to support the reforms that Yeltsin helped begin without banking all their assets on Yeltsin himself as the only embodiment of those reforms. Yeltsin's all-too erratic behaviour, and his tendency to retire from the scene in moments of crisis, have proved once again, if proof were needed, that dependence on specific personalities is a dangerous practice in international affairs.

Losing the Way in the Middle East

Yasser Arafat, too, has found it difficult to transform the abilities which helped create and sustain the Palestinian movement into those required to run a state. As the leader of a revolutionary cause, he lived and acted through secrecy and stealth. He gathered all decision-making powers to his own chest, and held them there closely. He controlled the funds, made the payments, provided the patronage. These operational methods stood him in good stead when he fought for his cause out of the shadows.

Once in the open, however, these same ingrained habits of governance have become an impediment. There are now too many decisions to be made for him to continue to refuse to delegate the smallest authority to anyone but himself. By insisting on placing his cronies in positions that require more than loyalty, he has alienated many whose technical expertise he needs, and has failed to build the kind of government that the situation requires. Many of the states which promised funds to help nurture the Palestinian authority established in the Oslo agreements, as well as the controllers at the International Monetary Fund (IMF), will not release those funds if Arafat does not establish greater transparency in the financial affairs of the incipient state.

The Palestinian Authority's consequent inability to fulfil any of the expectations that welled up when the Oslo agreements were signed have compounded Arafat's weaknesses. Radical Islamic groups within the territory Arafat nominally controls have grown stronger, threatening not only the validity of his position, but Israel's security as well. In the wake of suicide bomb attacks which killed up to 50 Israelis, Prime Minister Yitzhak Rabin's willingness to further the peace process dwindled to near zero.

Even before the shocking wave of deaths, Rabin had been reluctant to accept the need to move against the Israeli settlers who must be brought into line if the next stage called for by the Oslo agreements is to be reached. Under the best of circumstances this would be difficult, for many of the more ideological or religiously motivated would refuse to give up their homes, their schools and their places of worship without a battle. This would pit Israeli against Israeli and would require the firm support of most of those who have not settled in the Occupied Territories. In the climate created by the successful terrorist attacks, that support is lacking.

Rabin and the ruling Labour Party must face the electorate in mid-1996. He has watched dourly as his backing, and backing for his peace policies, have dwindled. Even moderates, who had been the most enthusiastic about moving away from confrontation, have been driven to ask difficult questions. President Ezer Weizmann, an early advocate of the policy Rabin chooses to follow, publicly voiced his doubts.

Both Rabin and Arafat are in a difficult position. Their constituencies have become more critical with each passing month. Yet their very weaknesses may be why they will finally agree further positive moves. Overcoming their own doubts and hesitations long enough to set in motion further steps in the process may, but only may, lift some of the negative pressures that have built up. Without some progress, however, it is clear that the political lives of both statesmen will soon be over. Aided by encouragement and support from the United States, which in the past two years has sent its Secretary of State on 11 trips of shuttle diplomacy to the region, recognition of this factor may help tip the balance. In the absence of strong leadership, it will have to do.

There's Still a Nuclear Threat

In one other area, that of nuclear-weapons policy, the United States has played a consistent and coherent role. Fourteen years after the beginning of negotiations and four years after they were successfully concluded, the Strategic Arms Reduction Talks Treaty (START I) finally entered into force at the end of December 1994. When the Treaty was signed in 1991 Washington had only Moscow as an interlocutor; in the intervening years the problem quadrupled. In three years, the US had to find the magic formula to ensure that Belarus, Kazakhstan and Ukraine, which acquired nuclear weapons on their territories when the Soviet Union ruptured, would voluntarily surrender them. Extensive and exhaustive diplomacy, resulting in security assurances, economic blandishments and promises of assistance in dismantling these weapons, provided the right results.

It will require the same kind of dedicated leadership and diplomacy to ensure ratification of START II. Because of the way the Russian and US forces are structured, the large reduction in force levels required by the Treaty would make it more difficult for Russia to accept than for the United States. For this reason, the Russian parliament has shown marked reluctance to accept the present Treaty. At their summit meeting in Washington in September 1994, Presidents Clinton and Yeltsin committed themselves to work for ratification by the time of the next summit and to examine possible follow-on agreements. Clinton will have an opportunity that he is certain to exploit when he visits Moscow for the 50th anniversary celebrations of the end of the Second World War. Whether Yeltsin will have sufficient influence over the Duma to make this an effective effort must be in some doubt.

This is an effort that must be made, however. The US has also taken the lead in the effort to find a large majority for the indefinite and unconditional extension of the Nuclear Non-Proliferation Treaty (NPT). The Treaty has been in force for 25 years and from mid-April to mid-May 1995 the 171 signatories will meet to determine its future. No world body of this size, dealing with a subject of this complexity, could be expected to show much consensus. And the permutations which have been discussed before the body has convened have been wide. They boil down, however, to whether the extension, which appears certain, is to be of definite or indefinite duration, and whether the many, mostly small, non-nuclear powers will be able to attach conditions forcing the few, but major, nuclear powers to accept conditions they do not want.

Early soundings suggest that the majority (only 86 votes are needed) would go for an indefinite extension without onerous conditions. But a simple majority would probably not carry the moral weight and feeling of universality that has helped make the NPT as effective as it has been. The Treaty requires this weight because it does not contain built-in mechanisms to stop rogue regimes from ignoring its principles even after they have accepted them. Continuing difficulties with North Korea have underscored this point.

After months of hard bargaining, the United States and North Korea announced in October 1994 that they had agreed a 'framework of steps' which, if implemented, would halt Pyongyang's efforts in the nuclear-weapons field. The relief that this brought was short lived, however. By early 1995 North Korea made it clear that it still sees agreements of this kind as merely the prelude to the search for greater concessions. As a major sweetener to entice Pyongyang's initial agreement, the US had promised to ensure that North Korea was supplied with two modern light-water reactors to replace the reactor it had to shut down to ensure that it would not produce plutonium to make bombs. The US had also convinced South Korea that the $4 billion that it would cost to supply such reactors was worth the peace of mind that would come from stopping the North Korean nuclear programme.

Pyongyang purported to be dumbfounded at 'suddenly' discovering that its arch-enemy was to get the credit for supplying such advanced plants. It seized on the pretext that it had not known where the plants were to have originated to indulge once again in brinkmanship. It insisted that if this impasse was not broken by 21 April 1995, the date on which it was to have signalled its acceptance of the reactors, it would withdraw from the agreement. In late March 1995, the US and North Korea were back around a table, with the US trying to salvage the agreement and Pyongyang trying to wring new bounty from an agreement that everyone else thought they had already made.

The East is not Immune

There are not many levers that the United States can pull to ensure North Korean compliance. The threat of sanctions against the recalcitrant regime is even more hollow this year than it was last. It would depend heavily on the support of Asian governments. None of them were eager then, and all of them are weaker now. It is also more difficult to judge what the effect would be upon the government in Pyongyang even if some sanctions plan was arrived at because there is some doubt as to who is in charge of that government. Kim Il Sung, the only leader North Korea had known in its 47-year history, died suddenly in July 1994. His ostensible successor is his son, Kim Jong Il, and there are propaganda indications that he may be stepping into his father's shoes. But at the end of March 1995 he had still not filled any of the formal positions of power which his father's death had left vacant. If there is a power struggle under way, it is very well hidden. It might be easier to deal with this inscrutable regime, however, if the outside world knew whether it was still dealing with the same leader, or faction, that had negotiated earlier agreements.

The role that China chooses to play is a key component in the resolution of this question, as it is for all East Asian security questions. Yet China, too, faces an imminent change of leadership. Deng Xiaoping, who has guided its domestic and foreign destinies, is over 90 years old, very ill, and clearly no longer in charge. His supposed successor, Jiang Zemin, has already been named and in theory is in command. There is good reason, however, to believe that the succeeding government will be more of a coalition than a one-man show. In any case it is a government facing more than its share of domestic social and economic problems; it has little desire to be embroiled in avoidable foreign issues, although it can be expected to continue to press for advantages where it perceives the cost to be minimal.

Weak government has not been the rule in Asia, but the region has caught up with the rest of the world. In South Korea, Kim Young Sam's popularity has dropped to low levels and he would prefer to deal with his own internal problems. The government in Japan is an extraordinary arrangement. Cobbled together by two parties at opposite ends of the political spectrum simply so that they could be in office, it married a Socialist Prime Minister with the conservative Liberal Democratic Party which had dominated Japanese political life for more than 40 years. It has managed to pass an anodyne budget and to complete the reform of the electoral system started by its predecessor. New elections are expected in summer 1995 and the faltering government will most likely be replaced. Until then it can hardly be expected to march to a bold new tune.

Asian governments generally approach security questions with trepidation and some timidity. Most of the societies in the region operate on a principle of consensus; decisions are taken only after some common agreement has been reached. Positive leadership is avoided, for it can lead to unhappy consequences. This is particularly true in their approach to security. Having mounted the need for consensus high on the totem pole, they then point to it in explanation for the slow approach to constructing a regional security network.

This method may work if there is not one in the region who insists on operating outside the boundaries that have been established, and then simply ignores complaints. The hollow nature of this implicit sort of contract was pointed up in early 1995 when China, which had said at a recent meeting of the ASEAN Regional Forum that it agreed with the 'consensus' that no one should use force to change the status quo in the Spratly Island chain, blithely occupied Mischief Reef which was claimed by the Philippines, a member of ASEAN. Having raised a principle to a dogma, no ASEAN voice was heard in protest. Unless the Asian nations begin to think in somewhat different terms about regional security, it is an unfortunate pattern that can be expected to recur, and often. At least in the West, the dispute in March 1995 between the Philippines and Singapore over the application of the death penalty to a Philippine worker in Singapore found guilty of murder showed up another potential weakness: public opinion in ASEAN countries will be a factor in international diplomacy which may in time make that diplomacy less cosy and, perhaps, less effective.

In Transition, but to What?

It strains credulity to believe that there are weak leaders in almost every state in the world simply because all the heroes are gone. In fact, in many countries the leaders are the same people who had previously been in power and were earlier able to take bold actions in pursuit of common goals. This is partly because it is more difficult to identify common goals when there is no longer a single powerful enemy to oppose. True enough, but the underlying reasons are far more complicated.

Leaders of states in the last years of the twentieth century are weak because the nation-state, as an institution, is weak. There has been a shift of problems from the national to the global arena. Transnational business, with its flow of monies ceaselessly and instantaneously across borders, cannot be easily regulated, nor countered from a single capital. Transnational crime has swamped the dams that national police forces have erected to protect their societies. The increased ease of transmission of information across borders, the ubiquitous and insistent television news screen and the development of the InterNet have all made governance more difficult. Environmental problems, the mass migration of the poor, no longer necessarily from just next door, and the free exchange of dangerous ideas have increased every government's burden.

These new problems are notable for their diffusion. Generally, strong leaders rise in response to a sharp crisis. A direct threat can be dealt with in a direct way: Franklin Roosevelt wrestled with a deep economic depression, Winston Churchill accepted and defeated Hitler's challenge, Harry Truman faced down Stalin's efforts to expand Soviet power. Some larger-than-life leaders reflect the firmness of an ideological vision. They strongly believe that they can rectify wrongs by doing things differently: Vladimir Lenin, Margaret Thatcher and even Ronald Reagan led from conviction. The amorphous challenges that have crept up in the present era are not easily countered, or conquered, by simple direct actions. Yet the only leaders in sight with vision and conviction are possessed by some form of fanatical ideology. For most, in these circumstances, muddling through is the only, even if uninspiring, style of leadership available.

Strategic Policy Issues

Fissile-Material Protection: A New Challenge

Of the numerous reports on fissile-material smuggling in 1994 two incidents were of particular significance. The interception of 5.6g of weapons-grade plutonium (99.75% Pu-239) mixed with 50g of metallic alloy in Tengen, Germany, in May; and the seizure of 2.72kg of highly enriched uranium (HEU at 87.7%), possibly from stockpiles of fuel for the Russian Navy, in the Czech Republic in December. For many Western officials these illegal movements of fissile materials were strong evidence of what Russian authorities were reluctant to concede – the inadequacy of safety and security systems for the protection of high-quality fissile material in the former Soviet Union (FSU).

The amounts of such material being offered for sale were insufficient to make a nuclear bomb, but a number of similar sales, if successful, could provide enough material for a device. Scientists continue to disagree over the level of technological sophistication needed to design a bomb using very small quantities of fissile material. It has been argued, for example, that a yield of 1 kiloton could be achieved with a 3kg plutonium core in a bomb similar to that tested by the US in 1945. Others believe that this amount is too small, but there seems little disagreement that the threshold set many years ago is too high.

Thus a proliferator choosing to purchase on the black market could accumulate sufficient fissile material to provide the core of a nuclear device much more quickly than previously thought possible. Given that 1kg of plutonium is only the size of a golf ball, once such small amounts leave a nuclear facility they are extremely difficult to detect. Thus it is essential that the material be continuously monitored, accounted for and kept under close safeguards at its source. How secure the former Soviet security system is has now become a major question. The efforts of the international community to ensure that there are no leaks suggests that it views Russia's controls with considerable concern. Much still has to be done before the world will be comfortable with the security of fissile material.

Fissile-Material Safety in the Former Soviet Union

The Soviet security system for nuclear installations relied largely on the 'human factor'. The emphasis was on perimeter defence rather than internal safety and security procedures. The absence of technical security and surveillance systems posed less of a problem when the civilian and military side of the nuclear industry was managed by one ministry (the Ministry of Atomic Energy – MINATOM), itself closely linked to the military. Both were directly subordinated to the highest echelons of authority. Military nuclear installations were located in secret, closed cities, shielded from the outside world by travel restrictions and perimeter fences. These factors combined to secure fissile material physically.

The command structure established during the Soviet era seems now to apply only to warheads mounted on their launchers. The Strategic Rocket Forces, responsible for maintaining and protecting missiles, are among the few units in the Russian military not affected by a loss of prestige and a deterioration in their internal cohesion. Unlike most of the armed forces, no reports of corruption have stained the honour of the Strategic Rocket Forces.

But in other areas the outlook is grimmer. Social transformations have weakened the old safety system, despite – some would say because – vast portions of the nuclear complex are still under MINATOM's authority. Nuclear installations outside Russia have now become the responsibility of other authorities in the newly independent states (NIS) of the FSU, but technical information often remains in Moscow. Closed cities are open and no longer secret. Some naval shipyards, which have been partly privatised, are now easily accessible from land or sea. Although accurate information on the degree of enrichment of Russian naval reactor fuel is still not available, it is estimated to range from 20% to 90%. At its higher level it is a desirable and accessible target.

Lines of accountability in both the armed forces and ministries have become blurred. The civilian industry, the Ministry of Internal Affairs, the KGB and the armed forces each had their separate systems. The material control and accounting system established in 1984 had always been shaped and executed by these different bodies. Apart from the fact that in most cases it only accounted for 'bundles' (fuel assemblies or containers), it has now virtually broken down.

A lack of opportunity for work, and extremely tight budgets, have combined to erode much of the matrix that ensured fissile-material security. The end of the nuclear build-up in the 1970s and 1980s has left Russian nuclear scientists facing redundancy. Yet the closed infrastructure of formerly secret cities offers little employment alternatives. The generally dire economic situation aggravates these structural changes.

MINATOM runs 151 enterprises. Some, such as enrichment facilities, were made economically independent and should now provide for their own survival. Yet many of the enrichment plants are stuck in a loop of debt, beginning with customers who fail to pay their electricity bills. This not only means that no money is available for safety improvements, but that workers are not paid for months. In these circumstances, a need for sheer survival, or an interest in personal profit, easily overrides administrative and social obligations. Reliance on the 'human factor' is undermined and can no longer provide a reliable shield against the illegal diversion of fissile material.

New National Safety Regulations

In December 1991, Russian President Boris Yeltsin created *Gosatomnadzor* (GAN), the State Committee for Nuclear and Radiation Safety. In principle, its remit was to design and implement safety regulations applicable to the military and civil activities of the nuclear sector and its enterprises. This included both the safety of material and export controls. MINATOM and the military were eager to preserve their monopolistic status, however, and despite presidential backing, GAN soon faced strong bureaucratic opposition, preventing it from exercising wide-ranging powers.

Since mid-1994, however, the need to improve nuclear safety has become clear to some in the Russian administration. GAN has therefore been given increased political support and at least limited access to military installations. A long-awaited law governing nuclear policy has had two readings in parliament and final approval is expected in the near future. In February 1995, Prime Minister Viktor Chernomyrdin convened a meeting to review national nuclear safety systems and ordered immediate improvements to be implemented.

In Ukraine the independent nuclear safety commission gained the respect of the West for the most effective effort in the NIS to regulate nuclear power development. In December 1994, however, President Leonid Kuchma merged its functions with the environmental ministry in a new Ministry for the Protection of the Environment and Nuclear Safety, headed by Yuri Kostenko, an outspoken critic of the Nuclear Non-Proliferation Treaty (NPT). As in Kazakhstan and Belarus, Ukraine still lacks the legal, administrative and technological infrastructure to establish a safety system that would meet Western standards. Nor is it clear whether enough political will exists to bring about the required improvements.

Multilateral Assistance

Foreign assistance to the FSU in the field of nuclear safety reflects the historical divide between relative openness and international supervision of civil nuclear programmes on the one hand, and national claims to secrecy in military applications on the other. International assistance largely builds on the US–Russian disarmament treaties, which provide a political and legal framework for cooperation in increasingly sensitive areas. For this reason activities have been divided roughly between US assistance to the FSU primarily in nuclear weapon-related areas, while other industrialised countries and organisations such as the International Atomic Energy Agency (IAEA) and the European Commission have concentrated on reactor safety and waste management.

In the latter area a large number of programmes have been launched, but they suffer from insufficient coordination, with competition developing as soon as national industrial interests come into play. In 1991 the European Union (EU) launched two assistance programmes for the East European countries (PHARE) and Technical Assistance to the Commonwealth of Independent States (CIS) (TACIS), particularly to Ukraine and Russia, building on work initiated by Sweden and Finland. Embedded in cooperative networks between EU institutions and their counterparts among administrators (MINATOM and GAN), research institutions (the Kurchatov Institute in Moscow) and reactor facilities, these programmes support information exchange, training (on-site and at special seminars), the provision of technology and twinnings between East and West European nuclear power plants.

The Concertation on European Regulatory Tasks (CONCERT), also launched in 1991 and supported by a number of Euratom agencies, is specifically aimed at establishing a 'pan-European' standard of nuclear safety regulations. CONCERT was expanded from a previously internal EU programme to the East European countries and the CIS. For the former it is part of their association treaties with the EU, intended to harmonise nuclear safety standards before they become members of the Union.

Although these EU activities are said to be coordinated with the IAEA, there is considerable duplication of effort between the two organisations. The Agency had stepped up its safety assessments of Soviet-built reactors in 1991. After the accession of Belarus, Kazakhstan and Ukraine to the NPT, it established bilateral cooperative relationships with the NIS. It coordinates programmes sponsored by European governments as well as by the Scandinavian countries, Canada, South Korea and Japan.

The objective of IAEA-coordinated programmes is to establish export controls and nuclear legislation, train nuclear engineers and operators in physical safeguards and accounting systems and improve the technical safety of nuclear installations. Although the IAEA has assessed the safety of the power reactors in Russia, it has not yet gained permission to implement a comparable programme of safety assistance there.

From a technical point of view the Agency feels that its efforts are bearing fruit. The NIS are entirely supportive, not least because they are themselves interested in assessing and ameliorating their own nuclear inheritance. Yet legislation has to be created from scratch. This is particularly time consuming because it depends on a clear allocation of responsibilities within national authorities. In addition, the technical requirements are so demanding that much more time, effort and money will have to be spent before Western safety standards can be achieved.

Norway and Japan in particular have been active for some years in offering support for Russian management of its low-level liquid radioactive waste from naval propulsion units. Japan has only recently added another agreement to provide technology for processing such nuclear waste, especially from submarines, in the Pacific region. Yet it has proved difficult even for government organisations, as well as for non-governmental ones, to gain access to the technical data required to implement these projects.

Some first steps for international cooperation to prevent nuclear transfers have also been taken. In August 1994, the Russian and German governments signed a Memorandum of Understanding in which they agreed to support the prosecution of cases of nuclear smuggling at the governmental, as well as law-enforcement, level. Similar bilateral agreements, which include the exchange of liaison officers and training assistance to police forces, have been signed with Ukraine and Germany's East European neighbours. The US Federal Bureau of Investigation (FBI), the Russian Interior Ministry and the German Federal Criminal Office further cooperate on information exchange and in trans-border investigations. Efforts to strengthen Europol, however, continue to run up against issues of sovereignty over national police forces.

US–Russian-Led Cooperation

An early basis for US–Russian cooperation in the field of fissile-material safety was laid with the Weapons Destruction and Non-Proliferation Agreement of July 1992, an umbrella agreement for the Cooperative Threat Reduction (CTR), or Nunn–Lugar, programme. This agreement provided the legal underpinnings for CTR funds to be spent in Russia. The US Department of Defense and Russia's MINATOM are the executing agencies for the nuclear-material projects developed under the Nunn–Lugar programme. This consists of other agreements – not only with Russia, but also with Belarus, Kazakhstan and Ukraine – and specific means for implementing individual projects.

The programme's often-criticised slow start is partly attributable to difficulties in communication both between officials and scientists of the two states, as well as within the former Soviet bureaucracies in defining and expressing their requirements. The primary aim of the Nunn–Lugar programme is the safe and secure dismantlement of nuclear weapons in the FSU. To this end it supports the safe transfer of missiles from the NIS to Russia and within Russia from deployment to storage sites, dismantlement, safeguards against proliferation, conversion of the defence industry, and alternative employment for former nuclear scientists.

The US Congress has now allocated about $1.2 billion to support this programme. Equipment, including tools for dismantling missiles, safe transport equipment such as armoured blankets, and radiation detection devices, are being delivered. Tenders have been invited for defence conversion and silo destruction projects drawing on local expertise. GAN has now also been added to the list of recipients entitled to financial and technical aid. This is primarily aimed at training inspectors and providing equipment for facility inspections. Longer-term projects envisage establishing a nationwide accounting system and support for safeguarding facilities which are either controlled by the Ministry of Defence or MINATOM. After considerable bureaucratic delay work is at last under way at Mayak for a facility to store plutonium from dismantled warheads. Russian authorities regarded this facility as the bottleneck in the smooth progression of disarmament, but seemed unable to overcome the obstacles that have repeatedly delayed the project.

In 1993 the Russian and US governments negotiated with the US Enrichment Corporation for the $11.9bn purchase of 500 metric tonnes of HEU from dismantled warheads. The HEU is converted to low-enriched uranium (LEU) in Russia and will be sent to the US over a period of 20 years. The US will sell the LEU on the open market. Although this has been delayed by problems over the required standard of specific isotope concentrations, it is significant in a number of ways. It guarantees that HEU will no longer be available for weapons production. It is also an important source of income for the Russian processing facilities.

An extensive programme for cooperation between nuclear-weapon laboratories entered a second phase in summer 1994. It is aimed at providing potentially jobless nuclear engineers and weapons designers with a future in alternative high-technology fields to prevent them from emigrating to potentially proliferating states. Argonne National Laboratory and the Russian ENTEK institute are cooperating on a study aimed at replacing HEU in Russian research reactors with LEU.

The teaming of the Californian Los Alamos weapons laboratory with a Russian counterpart, Arzamas-16, in July 1994 was one of the major events of this effort. In January 1995 the two facilities began cooperative research to develop physical safeguards for fissile material at the Russian site. A prototype security system based on a current US design will be installed at Arzamas-16 to allow for real-time accountability of all types of nuclear material. Its hardware consists of detection, monitoring and surveillance devices for nuclear materials and containers as well as portal monitors. Software for the computer network is based on the Los Alamos-developed US type. The plan is eventually to expand the project to other Russian facilities handling nuclear materials.

The International Science and Technology Center (ISTC) in Moscow supports alternative employment for scientists in a less specific way. The

ISTC is sponsored by the US, the EU and Japan, and recently by Finland. Within the broad framework of defence industry conversion the centre's main activities are in nuclear security, energy generation and environmental protection. A similar centre in Kiev, Ukraine, was founded in 1994 by the US, Canada and Sweden.

Reluctance on the part of Russian authorities to drop the barriers of secrecy is at least partly the reason why full fissile-material control became a funding priority only in 1994. In March, US Energy Secretary Hazel O'Leary and Viktor Michailov, head of MINATOM, signed a memorandum on the verification of nuclear warhead dismantlement which formed the basis for a three-step programme announced in the summer. Initial familiarisation visits to Russian and American storage facilities for plutonium removed from nuclear warheads led to a system of routine visits by the end of 1994. A second step is to develop coordinated methods for conducting reciprocal, non-intrusive, in-container inspection measurements on warhead plutonium. Negotiations on a scheme for the reciprocal confirmation of fissile-material inventories were begun in early 1995.

These steps were set in a wider framework by a further significant initiative in December 1994: the regular exchange of data on warhead dismantlement and stockpiles between the presidents of the US and the Russian Federation. The agreement depends on new security legislation to be passed by the Russian Duma, but the positive signal sent by the US Congress, which was prepared to change US national security legislation, should ease the way to parliamentary consent in Moscow.

Also in December 1994, O'Leary and Mikhailov, as co-chairs of the Nuclear Energy Committee, signed a formal agreement to exchange information on nuclear warhead and materials safety and security. This will cover the dismantlement process, technical means for the safety and security of nuclear warheads and materials and the development of criteria for releasing information on nuclear warhead designs to the press. Work on these matters began in January 1995.

On the practical level, a security and safety system at the Kurchatov Institute for the physical protection of fissile material has been put in place. Although its design was developed under US consultancy, the major part is based on Russian equipment. In addition, MINATOM and the US Department of Energy have recently made progress on a long-standing project for the joint inspections of six nuclear sites, including Obninsk and Elektrostahl. US authorities have long had grave concerns about the security and safety of fissile material kept at these sites.

Ongoing Problems

A number of problems remain. Although a US–Russian working group on plutonium disposition has now beeen established, there are fundamental differences in approaching the issue. While the US administration views plutonium as waste, Russia plans to use the material in fast reactors. Using the material as reactor fuel would be more accommodating in terms of non-proliferation, because it could not easily be reconverted for use in weapons. Yet breeder technology has been excluded from American nuclear reactor policy for more than a decade, because such reactors have the potential to produce rather than reduce plutonium.

The issue is of immediate importance, because although Russia has pledged to stop military plutonium production by 2000, MINATOM has declared it should do so only if a suitable alternative to the three remaining reactors at Tomsk-7 and Krasnoyarsk is found. Other than at Tomsk-7, where the options for a replacement are completely open, there is a half-built coal-fired thermal plant at Krasnoyarsk. It could be completed, but in August Yeltsin strongly expressed his support for a 'new "underground" nuclear plant'. One alternative, which may be acceptable to both sides, is MOX (mixed uranium-plutonium oxide) technology. MOX fuel can be used in thermal (slow) light-water reactors (LWRs), where the plutonium content of the initial load is reduced by one-quarter.

The US Academy of Sciences strongly advocated this option as an alternative to vitrification in a 1994 investigation of plutonium disposition. The US can already use MOX fuel in some reactors. In Russia, spending on new power plants cannot be avoided. Yet the application of breeder technology can. If Russia could produce MOX fuel assemblies, they could also be used in foreign LWRs, of which 250 are currently operated worldwide.

Politically most desirable are comprehensive agreements on verification of nuclear warhead dismantlement as well as fissile-material production. Given the unstable political situation in Russia and the forthcoming elections in 1996, it is important for the US and Russian governments to establish an irreversible legal framework which would bind future administrations to even greater transparency over fissile materials. Transparency lies at the heart of the issue in many ways.

The economic situation in Russia can be expected to stabilise only gradually. Incentives for illegal trade are expected to multiply. Greater transparency would demystify fissile material, thus presumably reducing its perceived exorbitant value on the black market. Increased transparency is furthermore likely to improve the access of other industrialised states to the technical information needed for an effective multinational effort. Better coordination of the various approaches of the donor countries would also help, and might even bring about immediate improvement. Duplication of effort is not only wasteful in financial terms, but also vastly increases the time necessary to complete all outstanding tasks.

There is a need for more open discussion of the nuclear issue, particularly in Russia and the NIS. Unlike the West, where the habit of democratic pressure on the legislature through public dialogue is well entrenched, the former communist states have still to adopt these practices. But national legislation on this question is vital. Once in place and with clearly defined lines of accountability, technical assistance can be provided in the right places much more effectively. Developing a complete nuclear safety culture would greatly enhance the effectiveness of the many programmes already under way. Open public debate could also have a positive effect on the development of better border controls within the CIS.

There are few more vital issues in national and international security than the control of material for weapons of mass destruction. In the long run this can only be accomplished by international cooperation. Both Russia and the United States must make certain that their ministries and electorates realise that the problem cannot be solved unless these former foes unify their efforts. Although a promising start has been made, it will require the same kind of dedicated effort that went into the construction of the nuclear threat to ensure its elimination.

Transnational Crime: A New Security Threat?

Organised crime has traditionally been seen as a law-and-order problem for individual states. But the growth of economic interdependence among states, the development of rapid transport and communications systems, the vast increase in international trade, and the emergence of a global financial market have dramatically changed the context within which organised crime operates. Not surprisingly, there has been an equally profound change in the nature of such crime itself.

Organised crime has taken advantage of new opportunities provided by globalism to become a transnational phenomenon that poses novel challenges to national and international security. The problem has been exacerbated by the emergence of a 'crisis of governance' both in particular states and in the state system as a whole. Political upheaval and dislocation has provided an environment in which criminal organisations not only can operate with relative impunity, but can also exploit new possibilities for economic enterprise.

The Traffic in Nuclear Materials

Nowhere has this been more obvious than in the former Soviet Union where the upsurge in organised crime has led some to conclude that the communist state is less likely to be replaced by a democratic state than by a one that is either dominated by organised crime or in which there is a symbiotic relationship between government and major criminal organisations. There are an estimated 5,700 criminal gangs in Russia with a total membership of over 100,000. While many of these are small and disorganised, others are more sophisticated and have links to the political and economic elite. Indeed, criminal organisations have infiltrated the banking industry to such a degree, and corruption in government is so extensive, that some now contend that transition to a market economy will result not in free competition, but in 'an economic system ruled by connections, bribery and extortion'.

What makes all this even more disconcerting is that Russia remains a nuclear superpower with not only a vast arsenal of nuclear weapons, but also much poorly inventoried and loosely controlled nuclear material in a wide variety of installations and facilities. The collapse of the Soviet Union has also resulted in the collapse of key elements of the security system in the nuclear industry. That workers in this sector are poorly paid and have lost considerable status since the end of the Cold War has created additional incentives to steal and sell nuclear material. When these conditions are juxtaposed with rising opportunities, it seems clear that the problem will get worse before it gets better.

Several seizures and arrests during 1994 highlighted the possibility that weapons-grade material might fall into the hands of nuclear terrorists, pariah states or organised criminals intent on large-scale extortion. Moreover, as the United States and Russia dismantle large numbers of tactical and strategic warheads, the amount of nuclear material that is available will increase.

Until 1994 nuclear trafficking seems to have been confined to fraudulent activities – especially the sale of 'red mercury' – and to non-weapons-grade materials. The discovery of enriched plutonium in Tengen, Germany, in May 1994, however, dispelled the complacency that had characterised many earlier commentaries on nuclear trafficking. The arrest in August of three men disembarking from a Moscow to Munich flight and the accompanying seizure of nuclear material underlined the dangers, as did a further incident in Prague in December.

The Tengen case was particularly disturbing because the discovery was accidental. Almost 6g of plutonium-239 were found at the home of businessman Adolf Jaekle, when law-enforcement authorities were searching for counterfeit material. It was subsequently revealed that Jaekle had widespread business connections, including links with former KGB and Stasi officers as well as with Kintex, a Bulgarian arms company with a reputation for questionable activities.

Had the Jaekle case been an isolated incident, the old complacency might have returned. On 10 August 1994, however, a Colombian dentist and two Spaniards were arrested at Munich airport in possession of 330g of enriched plutonium and 1kg of Lithium-6. Initially it was thought that the men might be linked to Colombian drug traffickers or to Basque terrorist organisations. Eventually, a much simpler explanation emerged: the three men all had financial problems and had been induced to bring the materials to Munich by German undercover agents. Even so, their success in obtaining a significant amount of high-grade plutonium was very sobering.

An even more serious incident occurred on 14 December, when 3kg of highly enriched (87%) uranium-235 were seized in Prague by Czech authorities. Three men who had worked in the nuclear industry were arrested – a Belarussian, a Ukrainian and a Czech nuclear physicist, Jaroslav Vagner, who had specialised in heavy-water reactors. Even though Vagner had not worked in the industry for several years, his involvement seemed to bear out fears that scientists and technicians would become involved in the illicit nuclear trade – and would be much more selective in their choice of material.

These were not the only incidents. Another episode involved a Turkish smuggling group, in which at least one member had been arrested previously for smuggling antiquities – suggesting that it is relatively easy for smugglers to switch from one commodity to another. According to a January 1995 report by the German Federal Criminal Office, the number of detected incidents rose to 267 in 1994. The percentage of fraudulent offers decreased by over 50% to 31.8%. 54% of offers came from non-Germans, most of them East Europeans, and there are signs of increasing links with organised crime. There was also at least one incident in which Russian police discovered nuclear material in the possession of a group of known criminals.

The extent to which Russian criminal organisations are involved in nuclear trafficking, however, remains uncertain. The major players seem to be former intelligence officers and unscrupulous businessmen, assisted by amateurs who know little about the material, equate radioactivity with high value and potential profit, and generally have to search for a buyer. Russian organised crime groups are involved on an opportunist basis, but nuclear trafficking is not one of their core activities. Other areas seem to offer higher pay-offs at lower risk.

It would be dangerous, however, to conclude that the peripheral nature of the business means consistent voluntary restraint. Organised crime groups have the capacity to acquire nuclear materials through coercing or bribing people within the industry – and will do so should they decide that there is a significant demand. As elsewhere, organised crime is enterprise crime and the product is secondary to the profit. So far, the predominance of evidence suggests that nuclear trafficking is mainly supply driven. Yet, the Jaekle case reveals that several more sophisticated supply networks may be operating and that these are organised, directly or indirectly, by potential end-user states such as Iraq or North Korea. It would certainly be a mistake to believe that there is no demand for weapons-grade material.

No other issue combines crime and security in quite the same dramatic way as trafficking in nuclear material. Yet the challenge to security posed by transnational organised crime does not depend only on a nuclear link. Transnational criminal organisations both exploit and exacerbate political instability and regional conflict. Criminal involvement in weapons trafficking has become a staple feature of the post-Cold War world. Such transactions can have significant impact both on local conflicts and on the capacity of terrorist groups to create mischief. It also seems to be the case that the drug trafficking and arms trafficking businesses are becoming increasingly intertwined with cases of ethnic involvement in the drug trade in order to purchase weapons. As one report noted:

> There is abundant evidence that organised crime is involved in the illegal arms trade and subversive activities that tamper with the rule of law in different parts of the world. The weight of the evidence indicates that it contributes to the political turmoil and upheaval occurring throughout the world. Drugs for weapons deals have become common in the world of organised crime, and many ethnic and political conflicts are aggravated by this unholy alliance.

Trafficking in People

One of the most consistent features of transnational criminal organisations is their engagement in what is essentially enterprise crime. The commodity they trade in is more or less irrelevant so long as it yields high profit and does so at acceptable risk. It is hardly surprising, therefore, that one aspect of transnational organised crime has involved trafficking in people. According to one informed estimate, criminal organisations are now attempting to smuggle as many as one million people a year from poor to wealthier countries.

Not only does this threaten a basic ingredient of national sovereignty, it also places in jeopardy the immigrants themselves, who sometimes end up in the poorer countries of Eastern Europe, Latin America and Asia rather than their desired destinations in Western Europe and the United States. Moreover, the would-be immigrants are highly vulnerable, and women in particular are often forced into sexual slavery. Even those who arrive at their destination usually owe money and are forced to engage in criminal activities for several years to pay off the debt.

Although this problem has received a great deal of attention in Washington, largely in response to an increase in the number of Chinese and Mexican illegal immigrants, it is not a problem confined to the United States or limited to Chinese and Mexicans. According to some estimates, immigrants from China are only 20% of those in a pipeline that encompasses India, Iran, Iraq, Pakistan, Romania, Sri Lanka and Sudan. Those in transit include 60,000 Chinese in Moscow, 80,000 or more Asians, Africans and Middle Easterners in Romania, and significant numbers in staging areas such as Spain, the Netherlands and Guatemala.

The scale of the problem is also revealed by the fact that in 1993 there were 1.2m attempts to enter the United States illegally. Moreover, smuggling organisations offer substantial bribes to officials and are ingenious in developing new smuggling routes. One estimate has suggested that the trade yields an annual profit in the region of $3.5bn. For the destination states and sometimes the transit states, however, the trade imposes significant and unexpected costs in terms of medical care, food, housing and transport for the illegal aliens. The problem has also begun to provoke a backlash against legal immigrants and legitimate refugees.

Trafficking in Drugs

There has been considerable speculation that some of the organisations engaged in alien smuggling are also active in drug trafficking. The infrastructure and routes are certainly very similar. If smuggling in people yields significant profits, however, these pale in comparison with those for illicit drug trafficking. Although estimates of the value of the global drugs trade vary considerably, it may be second only to the global trade in oil.

Moreover, the drug trade is still expanding both geographically and in terms of its products. Several important trends in the drug-trafficking industry have become evident in the last few years. One of the most important has been the great increase in the supply of heroin. Myanmar (Burma) and Afghanistan are the two major opium-growing countries, and Pakistan has emerged as a major processor of Afghan heroin. Turkish and Nigerian criminal organisations are also heavily involved in heroin trafficking, and the extensive activities of the Nigerians (in cocaine as well as heroin) have contributed to a rise in drug abuse throughout southern Africa.

The Colombian cocaine cartels have also diversified into heroin, and there is some evidence that heroin use is on the upsurge in the US. New and purer forms of heroin can be smoked rather than injected. Given concerns over AIDS, this is an important marketing consideration. As well as diversifying products, the Colombians have also diversified markets. Significantly, the Cali cartel – partly through its links with the Sicilian Mafia – has helped to create a large and lucrative cocaine market in Europe, whose profits may now exceed those from cocaine sales in the US.

More and more states are being drawn into the orbit of drug traffickers and transnational criminal organisations. Brazil, for example, is playing an increasingly important role as a transhipment state and has developed an extensive drug consumption problem of its own in the *favelas* of Rio de Janeiro. The police have also become so corrupted by the trade that the Brazilian authorities, in late 1994, felt compelled to send in the military in an effort to restore some semblance of law and order to the slums.

In many respects, the corruption of the police in Rio is typical of a broader problem that faces most states in which transnational criminal organisations operate extensively – the corrupting influence of illicit money. This tends to be particularly pervasive in the home states of the criminal organisations. In order to ensure that the environment is relatively hospitable and risk-free, the organisations engage in extensive corruption of local and central government and the judicial establishment. For government officials in developing states in particular, the financial rewards that can come from simply ignoring – let alone facilitating – the activities of transnational organised crime vastly outweigh those attainable through legal means. And since the alternative may be violence or death, the temptation to accept corruption as a way of life is profound.

Trafficking in Money

Transnational criminal organisations also pose a threat to the integrity of national financial and commercial institutions. Infiltration of licit institutions, intimidation of their owners, and distortion of their purposes to serve criminal objectives can all considerably augment the power of criminal organisations. To the extent that these organisations are deeply embedded in the legal economy and legal business, they are more resistant to law-enforcement efforts aimed at their disruption and dismantlement. Symbiotic relationships with legal business and the political elite – as in Japan and Italy – give criminal organisations additional protection. And the vast profits they make facilitate the development of such relationships. It is possible that similar symbiotic relationships will develop in Russia as the criminal and government sectors each mature, thus making the problem more one that resembles that in Japan and Italy than the problem that exists in Colombia.

Some analysts also believe that transnational organised crime poses a threat to the integrity of global financial institutions and mechanisms. Money laundering has become a major business in its own right, with criminals taking advantage of bank secrecy laws and offshore tax havens in places such as the Cayman Islands, as well as the electronic capabilities that facilitate the rapid transfer of funds around the global financial system. If money laundering further erodes the ability of governments to manage, control and regulate the global financial system, however, it does so only at the margin. Even though the amount of laundered money is staggering in absolute terms, in relative terms it remains a very small percentage of the money circulating in the global financial markets.

The capacity of criminal organisations to disrupt the system, therefore, should not be exaggerated. Nor is it clear that criminal organisations have any incentive for disruptive actions. Generally, they like the global financial system: it offers them opportunities to launder and invest their profits with relative ease. Nevertheless, it is conceivable that by moving their money out of a particular institution at a key moment they could cause localised problems. Although there was no organised crime involvement, the collapse of the UK bank Barings in February 1995 highlights the vulnerability of even venerable financial institutions to the rapid withdrawal of funds and the impact of disruptive and unexpected events on the stability of financial markets.

Perhaps even more disturbing than the vulnerability of the global financial system is the vulnerability of information and communication systems. Transnational criminal organisations are likely to evolve in ways that enable them to exploit new opportunities resulting from the development of highly sophisticated communication systems. One assessment of the nature of organised crime in the twenty-first century, for example, has suggested that the computer hacker will be an indispensable member of any serious criminal organisation. In this connection, it is worth noting that the Pentagon is placing considerable emphasis on both the offensive and defensive components of information warfare. Although the primary emphasis is on information warfare during hostilities, it is also clear that there are many peacetime applications of the capacity to destroy or disrupt information and communication systems.

The growing importance of information technology may eventually lead to a shift away from the traditional paradigm of security. Until now, national security has usually focused on threats to territorial integrity. While these traditional threats are unlikely to disappear, new and asymmetric vulnerabilities are emerging, vulnerabilities that could all too easily be exploited by transnational criminal organisations. As modern societies become more dependent on sophisticated communication and information systems, the possibility that these systems will be compromised or disrupted becomes more salient. Transnational criminal organisations can develop the capability to inflict damage of this kind relatively easily; to the extent that they feel threatened by law-enforcement efforts, they are also likely to do so. Indeed, the combination of the drug cartel or organised crime group and the computer hacker could prove both dangerous and intractable. Sophisticated hackers are not only notoriously difficult to trace and catch, but also represent a new form of individual empowerment that could have far-reaching and damaging consequences.

Security Challenges

The threat from transnational crime and criminal organisations is insidious, pervasive and multifaceted. If some of the threats are novel, many of them strike at the heart of individual states and in some respects at the state system itself. Transnational crime inescapably violates national borders, thus challenging one of the most fundamental attributes of sovereignty. Similarly, although the main aim of transnational crime is profit, the inevitable by-product is a generally implicit, but sometimes explicit, challenge to state authority: not to the military strength of the state, but to the prerogatives and powers that are an integral part of statehood.

While states in transition and developing states are particularly vulnerable to these threats, no state is immune. Transnational criminal organisations, by their very nature, undermine civil society, add a degree of
turbulence to domestic politics, and challenge the normal functioning of
government and law. While they are particularly effective where government is already weak or unstable, criminal organisations add further lay-
ers of instability.

They have much the same impact at the international level. Attempts
to regulate the global political system and establish codes of conduct,
principles of restraint and responsibility, and norms of behaviour are
increasingly challenged by transnational organised crime. Regimes to inhibit the proliferation of weapons of mass destruction are highly dependent upon cooperation among suppliers and the ability to isolate rogue
states that seek to obtain such capabilities. To the extent that pariah states
are able to develop alliances of convenience with transnational criminal
organisations, they will be able to circumvent these restrictions and erode
the effectiveness of the regime. Given the range of activities and the variety of threats posed by transnational organised crime, it is essential that
governments initiate more stringent measures and engage in more comprehensive cooperation designed to contain, disrupt and ultimately destroy these organisations.

During 1994 there was a growing recognition of this requirement. In
the US, President Bill Clinton categorised international crime as a threat to
national security. Perhaps even more significant was the World Ministerial
Conference on Organised Transnational Crime, held in Naples from 21 to
23 November. This conference was convened by the United Nations (UN)
to assess the dangers posed by transnational organised crime and to identify various forms of international cooperation for its prevention and control. Bringing together ministers of justice and ministers of the interior, the
Conference concluded with a Political Declaration and Action Plan designed to initiate more effective measures to prevent and control cross-
border criminal activities.

The key components of this action plan included the acquisition of
better knowledge about transnational organised crime, the need for substantive legislation at the national level to impose penalties for participating in criminal associations and conspiracies, and the development of
better evidence-gathering and witness-protection schemes. It was further
agreed that these measures should be accompanied by bilateral and multilateral cooperative arrangements, such as provision for extradition and
mutual legal assistance treaties, and greater efforts to control money laundering and confiscate the proceeds of crime. The action plan also included
an agreement to study the merits of a convention against transnational
organised crime, along the lines of the global convention against drug
trafficking. That it did not include agreement on such a convention re-

vealed that the international community has yet to develop a comprehensive and effective strategy against this scourge.

Despite its failings, the Ministerial Conference was a major step forward in recognising that transnational organised crime has become a global problem and one that jeopardises international security and stability. That it is a threat is no longer in doubt. How much of a threat will depend in large part upon the commitment of governments to deal with it. Success will require states to devote far more resources than ever before to such things as judicial assistance to ensure that there are few risk-free sanctuaries for transnational criminal organisations.

Resources alone are not enough. Governments also have to be more innovative and display greater ingenuity in the way they use their resources. Expertise that combines national security intelligence and law-enforcement information and methods is essential. But even if governments are more committed and more imaginative, this is no guarantee of success. The dialectic between law enforcement and organised crime is a continuing one of measures and countermeasures. Innovation in law enforcement is often negated by the capacity of criminal organisations to adapt very rapidly. Network organisations are both highly flexible and remarkably resilient. The implication is that the threat to national and international security from transnational organised crime is likely to be an enduring one.

The Global Spread of Dual-Use Technology

The international spread of commercially available technologies with potentially significant military applications threatens to complicate regional military balances and international security. Some advanced technologies, such as remote sensing satellites and global positioning systems, offer both important civilian and military applications. But with the Cold War over, technologically advanced countries are focusing more on sustaining their industrial bases and remaining competitive in global markets than on controlling the global spread of dual-use technologies.

Such technologies can be applied to a broad range of military tasks, including surveillance, navigation, targeting precision weaponry, and command, control and communications. The overall effectiveness of a nation's military forces can be substantially enhanced by these critical supporting technologies. The global spread of dual-use technologies raises several key questions: why is there an increasing diffusion of these technologies in the post-Cold War era? How might the growing commercial availability of dual-use weapons affect regional stability and international security? And how can the potentially negative effects of such advanced technologies be limited? Two key technologies, higher-resolution civilian imaging satellites and relatively precise spaced-based navigational systems, have acquired growing prominence since the 1991 Gulf War and epitomise the problem. Both of these demonstrate how the information revolution is steadily eroding any clear distinction between civilian and military applications in certain technological areas.

Lifting the Barriers to Technology Diffusion

With the end of the Cold War, several trends involving international politics, economics and technological development are reducing the barriers to the global diffusion of dual-use technologies. The dissolution of the Soviet Union has allowed Western allies to relax internal controls and to disband in March 1994 the Coordinating Committee for Multilateral Export Controls (COCOM), a key mechanism for limiting international access to dual-use technologies. At the same time, the demand for basic technologies with military applications is rising among newly industrialised countries. To guard against uncertain regional security developments, many nations in Asia and elsewhere are upgrading their own defence capabilities not only by purchasing foreign weapon systems, but also by acquiring the basic technologies needed to produce and support such armaments at home.

Economic considerations give added impetus to the spread of dual-use technologies. Facing major reductions in military budgets, some US, Russian and European defence firms are trying to break into civilian technology markets. Securing a competitive position in the global market for high technologies is a top priority for most advanced countries. Even many newly industrialising nations are seeking to develop their civilian and military technology.

A basic transformation is also under way in the relationship between civilian and military technologies. Through much of the post-war period national leaders could justify large investments in military research and development as yielding certain 'spin-off' benefits for the civilian economy. More recently, however, the civilian sector has become the undisputed source of advanced technologies in such essential fields as information processing and communications. The civilian economy increasingly provides a 'spin-on' benefit by producing cutting-edge technologies that are being adapted for military use. Many countries can bolster their military potential by improving their technology base in areas such as microelectronics, simulation capabilities and communication networks.

Space-based Navigational Systems

Space-based navigational systems demonstrate how broadly and rapidly a dual-use technology with revolutionary civilian and military applications can spread. During the 1991 Gulf War, highly accurate navigation data were used to support modern military operations and precision-guided weapon systems. At the same time, navigational data provided by satellite systems are becoming increasingly integral to certain civilian uses.

The US Navstar Global Positioning System (GPS) is a constellation of 24 satellites providing US military users with access to precise, three-dimensional positioning data. By using equipment that receives and interprets the electronic signals being transmitted by several satellites, the user can obtain local information accurate to 16 metres. Such data are important to military missions, including mapping, target acquisition and mission-route planning for aircraft and cruise missiles. The Russians are fielding a roughly comparable navigation satellite constellation known as GLONASS.

To permit separate military and civilian applications, the developers of the GPS system incorporated a selective availability feature that reserves the most precise navigational data for authorised military users. Data whose accuracy has been degraded to a range of 100 metres, would be available for worldwide civil use. But sophisticated commercial receivers that enable any user to acquire precision data from the GPS satellite transmissions are increasingly available. For example, Differential GPS technology can restore much of the positional accuracy by adding ground-based reference stations. A number of civilian agencies plan to take advantage of the Differential GPS technology, including the US Federal Aviation Administration, which is using this technology to improve the safety of commercial air navigation. There is also substantial international interest in applying GPS technology to facilitating road and rail travel.

The military utility of space-based navigational systems was highlighted in the Gulf War as GPS data guided the coalition forces' manoeuvre units through the mostly featureless deserts, provided launch coordinates for sea-launched cruise missiles, established the precise locations of artillery and air-defence units, and helped to guide aircraft and their weapon systems towards Iraqi targets. The commercial availability of this technologically advanced equipment was also clearly demonstrated. Because they did not have enough military GPS receivers, the US forces made up the difference by buying small, lightweight commercial receivers, which offer about 25m accuracy. Given the coalition forces' heavy reliance on commercial receivers (over 80%), the GPS system's selective availability feature was turned off in order not to degrade the navigational data for units lacking the military receivers. In light of the Gulf War experience, the United States is developing GPS receivers that are embedded into the regular inertial navigation systems (INS) used by military aircraft, helicopters and other vehicles. The United States and United Kingdom are even testing how the GPS system might be adapted to improve the accuracy of artillery shells.

The ability to exploit the military potential of these space-based navigational systems is not limited to the leading military powers. Third World countries might use GPS and GLONASS to develop new types of cruise or ballistic missiles. Existing inertial guidance systems can be adapted to receive updates from navigation satellites. An integrated INS/GPS guidance system can be coupled with a relatively inexpensive airframe, such as those provided by unmanned airborne vehicles, to produce relatively inexpensive cruise missiles capable of accurately delivering lethal payloads against enemy targets at long ranges.

Civilian Imaging Satellites

The growing commercial availability of higher-resolution satellite imagery data and technologies also highlights how the global spread of dual-use technologies is blurring the distinction between civil and military applications. During the Cold War, the imagery data produced by civilian systems offered substantially less quality and timeliness when compared with US and Russian military reconnaissance satellites. But several advanced civilian imaging satellite systems, scheduled to become operational in the next few years, will offer unprecedented international access to higher-resolution imagery data.

The US and Russia are greatly relaxing their tight internal restrictions on the civilian applications of imaging satellite data and technologies. New commercial enterprises are emerging as an important means of sustaining the skills and infrastructure underpinning military and intelligence satellite programmes in a time of substantially reduced defence expenditures. Furthermore, many experts believe that a robust market exists for higher-quality imagery products, supporting services and, in some cases, even the export of advanced remote sensing systems. The 'value added' industries involved in geographic information systems that process, analyse and distribute imagery data are estimated to have particularly strong commercial growth prospects.

National Programmes and Plans

Several new or upgraded civilian imaging satellites are scheduled to begin operating over the next five years (see Table). Compared with current satellite operations, such as the US *Landsat* and France's *Système Pour l'Observation de la Terre (SPOT)*, the next generation of surface remote sensing systems will offer much higher spatial resolution for distinguishing ground objects. The highest resolution is associated with panchromatic photographs from satellites carrying advanced electro-optical imaging sensors. In addition, most of these new satellites will carry multispectral sensors capable of collecting data of moderate resolution (about 4–20m) concerning the Earth's surface features. Multispectral data are essential for performing traditional remote sensing tasks such as crop forecasting, urban planning and environmental change assessments. Some other remote sensing satellites use synthetic aperture radar, which provides all-weather coverage, as their main sensor. Some new civilian imaging satellites will have special features, such as the ability to produce stereo images, necessary for accurately measuring the heights of objects and producing three-dimensional images, or the ability to disseminate imagery data more rapidly to the user.

Table 1 *Selected Civilian Imaging Satellite Programmes*

Name	Country	First Operational	Sensor Type	Resolution (m)
Landsat-5	US	1985	multispectral	30
Resurs-F or military imagery	Russia	1992–94	panchromatic	2–10
SPOT-3	France	1993	multispectral; panchromatic	20 10
Earlybird	US (World View)	1996	multispectral; panchromatic	15 3
Space Imaging	US (Lockheed)	1997	multispectral; panchromatic	4 1
Eyeglass International	US	1997	panchromatic	1
Quickbird	US (Ball Corp.)	1997	multispectral; panchromatic	4 1
SPOT-5	France	1999–2000	multispectral; panchromatic	20 5

Note: Excludes several remote sensing satellites with surface resolution capabilities over 5m, including the ALMAZ (Russia), JERS-1 and ADEOS (Japan), IRS-IC (India) and Radarsat (Canada).

Sources: Guide to Space Issues for the 1990s (Los Alamos, CA: Center for National Security Studies, Los Alamos National Laboratory, December 1992); US Office of Technology Assessment, *Remote Sensing Data: Technology, Management, and Markets* (Washington DC: USGPO, 1994); and Berner, Lanphier and Associates, Inc. (Bethesda, MD).

Internal US barriers to substantially improved civilian imaging satellites were considerably lowered in March 1994 when President Clinton signed Presidential Decision Directive 23 (PDD-23). This directive marked a major shift in US technology export policy. It permits private firms to operate commercial remote sensing systems and to sell images produced by space-based sensors to foreign customers. The new policy also kept open the possibility that proposals by American firms to export remote systems abroad would be considered on a case-by-case basis. Although the Clinton policy imposes certain conditions on US firms seeking licences for private remote sensing systems, few restrictions are placed on peacetime imaging operations. It now takes a high-level decision, either by the President or several top cabinet members, to restrict how firms approved for remote sensing operations can collect or distribute commercial satellite imagery data.

Similarly, the new US policy does not specifically limit the critical design features of the planned commercial imaging projects. Thus, some of them, such as the Eyeglass International and the Space Imaging Satellite, will be able to collect imagery data at 1m resolution. The White House apparently accepted the argument that higher-resolution imaging data would be increasingly available on the international market from foreign remote sensing enterprises regardless of whether image quality restrictions were imposed on US commercial satellites.

France's series of *SPOT* satellites has set the standard for commercial imaging operations in recent years. The *SPOT*-5 system, which is not scheduled for launch until the end of the century, is projected to have a 5m resolution capability. President Clinton's decision, therefore, permits US firms to develop new imaging satellites capable of producing higher-resolution imagery data than existing or currently planned European remote sensing satellites. Washington's new policy has provoked strong criticism from Paris. French officials have repeatedly disclaimed any intention to sell civilian imagery data with resolutions exceeding 5m because of the potential military utility of such data. The US policy change not only creates strong commercial competitors to the *SPOT* imaging operations, but also complicates France's effort to attract other European states, such as Germany and Italy, as investment partners in its *Hélios* military imaging satellite programme. Thus, the growing availability of commercial satellites offering relatively higher-resolution imaging data could erode the chances of realising an independent European military satellite capability.

Russia has already set the precedent of selling high-resolution imagery data. Struggling Russian space system enterprises are selling imagery data that apparently come from military reconnaissance satellites capable of very high resolution. This panchromatic imagery data appear to be purposefully degraded to approximate resolutions in the 2–5m range. Some reports suggest that Moscow is even planning to sell imagery that can resolve ground items that are less than 1m.

Military Value of Civilian Imaging Satellites

The demand for improved civilian satellite imaging systems was substantially heightened by the Gulf War which highlighted the military benefits of access to space-based imaging systems. Iraq lacked important intelligence about key coalition force activities during *Operation Desert Storm* because it was denied access to both military and civilian imagery. For example, in the month prior to the coalition force's surprise flanking manoeuvre that drove directly into Iraq the allies were able to reposition some 270,000 troops westwards without being detected. Conversely, the allies used their access to the civilian imaging satellites to update maps of the conflict area and to help plan specific military operations.

The higher-resolution imagery data offered by the new civilian imaging satellites will greatly enhance their potential role in supporting military and intelligence applications. The Gulf War demonstrated that the lower-resolution imagery data provided by *Landsat* and *SPOT* had military value. Higher-quality data will be even more useful in planning military operations and in pinpointing facilities for targeting purposes. Thus, the next generation of civilian satellites will provide a valuable source of imaging data to many military planners around the world.

The imagery provided by commercial operations still suffers in comparison with the data produced by dedicated military reconnaissance satellites. It is difficult when using civilian operations to obtain the timely imagery data that military planners require. More important, however, is the question of assured access. Foreign users of commercial imagery run the risk of their access being cut off at the source if they are suspected of using the imagery data for military purposes during a crisis or conflict. Thus, despite the improved quality of imagery data offered by new civilian imaging satellites, many countries will still seek to acquire military imaging satellites or to expand their existing reconnaissance capabilities.

Security Implications

The strategic importance of dual-use technologies is rooted in the information explosion that confronts national leaders and their defence planners. Along with precision weaponry and computer simulation capabilities, these information-intensive technologies are hailed by some experts as the technological precursor to a new military-technical revolution. If combined with innovative doctrines and organisational structures, such technologies could dramatically advance military effectiveness, particularly in identifying, fixing and precisely destroying numerous targets with highly lethal conventional munitions at long ranges.

The Gulf War is often cited as evidence of the strategic edge that can be gained by achieving 'information dominance'. The coalition forces enjoyed clear superiority in space-based technologies for reconnaissance, navigation, command, control and communications, and target acquisition. The crushing military defeat suffered by Iraq, despite its formidable arsenal of conventional armaments, demonstrates the perils of confronting adversaries who are able effectively to wage information warfare.

Within this context, the global spread of dual-use technologies has mixed implications for the post-Cold War security environment. Regional stability could be enhanced in some ways by commercially available technologies; higher-resolution imagery data and precision navigational data improve a nation's defensive capabilities. Access to such advanced technologies can strengthen the defensive military capabilities of local powers by reducing their vulnerability to foreign military threats and diplomatic coercion. Similarly, civilian imaging satellites could be used to foster regional transparency in the military activities of potential adversaries. Such transparency measures might be part of a broader regional security regime that combines commercially available satellite imagery with other types of confidence- and security-building measures, such as data exchanges.

On the other hand, regional instabilities might be exacerbated by the unrestrained proliferation of certain dual-use technologies. For example, a local power might try to alter the existing regional military balance by using advanced technologies to develop a non-nuclear strike capability against high-value targets in neighbouring countries. Such efforts could give rise to pressures for pre-emptive attacks among regional antagonists in times of crisis. Similarly, attempts by regional adversaries to gain a military edge over their rivals are likely to fuel bilateral or multilateral arms competition in peacetime.

Local powers might also use dual-use technologies to offset external pressure from a major power or even the international community. A likely objective would be to acquire technologies that diminish an external power's ability to threaten or project military power against them. Commercially available positioning equipment and satellite imagery could help a weaker state develop a more credible capability for increasing the costs, particularly in terms of casualties, that external powers must consider in deciding whether or not to intervene in a regional conflict.

Nonetheless, the prospects for containing the global spread of dual-use technologies are limited by strong international demand for advanced technologies and economic pressures on the supplier nations. The technologically advanced countries are ambivalent about restricting dual-use exports given the end of the Cold War and the growing prominence of economic competition in shaping national policies. Export controls on technologies that are critical to the development of weapons of mass destruction have received renewed attention from the United States and other supplier countries, particularly in the aftermath of the Gulf War. PDD-23, for example, requires a country discovered using the technologies for military purposes to be immediately cut off from information. But such countries can turn to other commercial suppliers to supplement the information received from the United States. And the efforts to create a more inclusive coordinating body for restricting the proliferation of other dual-use technologies to replace COCOM have yet to succeed. Meanwhile, the number of countries with access to militarily significant dual-use technologies will indiscriminately grow as short-term commercial considerations overshadow long-term global security concerns.

A Quiet Year for Arms Control

Although arms-control efforts were largely out of the public eye during 1994, considerable progress was made in resolving outstanding issues. Much was accomplished, but much remains to be done. Particularly successful was the implementation of the Lisbon Protocol by the United States, Russia, Ukraine, Belarus and Kazakhstan. These five states had negotiated the Protocol in May 1992 as a means of ensuring that no new nuclear-weapon states would emerge from the break-up of the Soviet Union. Lisbon made these five countries parties to the Strategic Arms Reduction Talks (START I) Treaty, with the proviso that the START I reductions would remove all nuclear weapons from Ukraine, Kazakhstan and Belarus. These three states would at the same time agree to become non-nuclear-weapon states under the NPT.

Following conclusion of the Lisbon Protocol, the denuclearisation of Ukraine, Kazakhstan and Belarus required extensive diplomacy as these countries negotiated for security assurances and assistance in dismantling the weapons systems on their territories. The Clinton administration worked closely with each of them, and with Russia and the UK, to develop acceptable assurances. The administration also developed an extensive programme of denuclearisation assistance for each. As a result of these efforts, the five Lisbon Protocol signatories brought START I into force in December 1994, 14 years after negotiations first started, and nearly four years after the Treaty was signed.

Months of tense negotiations between the United States and North Korea succeeded in October when an 'agreed framework of steps' was signed that is intended to halt and eventually reverse Pyongyang's nuclear ambitions. Agreement was also reached by the parties to the Biological and Toxic Weapons Convention (BWC) to negotiate legally binding measures to enhance confidence in compliance with the agreement's provisions.

In other areas, however, progress proved more difficult. Multilateral negotiations on a comprehensive nuclear test ban stalled over many issues, including the scope of banned activities, verification and conditions for the treaty's eventual entry into force. While it had received unanimous support in the UN in 1993, a permanent halt to the production of fissile materials for weapons purposes remained a distant prospect, with countries responsible for concluding this convention failing to agree even on a mandate for negotiations. And most importantly, the Chemical Weapons Convention (CWC), which bans chemical weapons worldwide and has been signed by 159 countries, did not enter into force in January 1995 because too few signatories had moved to formal ratification.

If 1994 proved to be a quiet year for arms control, 1995 may turn out to be its opposite. Bilateral negotiations between the United States and major countries of the former Soviet Union (notably Russia) may become more difficult. Ratification and rapid implementation of START II is next on the agenda, but this may fall victim to the vicissitudes of changing domestic political circumstances in Washington and Moscow. Progress on further denuclearisation, which has thus far proceeded with relative harmony, may also become more difficult. As for the global non-proliferation agenda, 1995 may prove a crucial year. The NPT will not only come under substantive review, but a decision must also be made on its duration. Current discord between the nuclear powers and their allies on the one hand, and much of the developing world on the other, does not bode well for the Treaty's long-term, let alone indefinite, extension. And failure to gain ratification of the CWC by 65 countries in 1995 may well doom this treaty for the foreseeable future.

The United States and the Former Soviet Union

Three issues dominated the arms-control agenda between the United States and the newly independent states of the FSU: the START treaties which reduced US and Russian strategic nuclear forces; cooperative denuclearisation efforts, particularly in relation to controlling fissile materials; and clarifying the Anti-Ballistic Missile (ABM) Treaty's provisions regarding non-strategic missile defences. Intensive, expert-level talks produced much progress, particularly in regard to START and fissile materials, but major issues still remained to be resolved.

Strategic Arms Reductions

In return for guarantees by the UK, Russia and the United States and shipments of nuclear fuel for its reactors, Kiev agreed to return all nuclear weapons on Ukrainian territory to Russia by the end of 1996. On the basis of this agreement, the Ukrainian Rada ratified START I in March 1994 and voted to accede to the NPT in November, a central US and Russian condition for activating the START Treaty. Instruments of ratification were exchanged the following months among the five Treaty parties – Belarus, Kazakhstan, Russia, Ukraine and the United States.

With the entry into force of START I and with many of its provisions already implemented unilaterally by the parties, ratification of the START II Treaty could formally proceed. Opposition within the Russian Duma to START II, however, has been heavy ever since its conclusion in January 1993. The Treaty reduces US and Russian nuclear forces to 3,000–3,500 weapons and requires the elimination of multi-warhead, land-based missiles – long the backbone of Soviet/Russian strategic forces. To conform with the Treaty's provisions and not to abandon the politically important position of numerical parity, Russia would either have extensively to restructure its forces by emphasising sea-based over land-based missiles or be compelled to embark on an extensive land-based missile modernisation programme. Neither of these options is particularly welcome at a time of great financial stringency. On the other hand, the START II provisions fit nicely into current US force structure plans. Indeed, the Pentagon's much-vaunted Nuclear Posture Review, released in September 1994, prescribed a force posture consistent with the Treaty's provisions that required little, if any, modifications in future plans.

The future of START II is thus uncertain. The Duma's concerns might be alleviated were the United States willing to consider a follow-on agreement with still lower force ceilings, a step that would ease Russia's dilemma while presenting new choices on the future shape of the US force posture. Prospects for a move in this direction in 1995 would appear to be slight, however. President Clinton formally endorsed the force posture recommendations contained in the Pentagon's Nuclear Posture Review, which bars reductions below START II levels until that Treaty is fully implemented. Moreover, the new Republican Congress is expected to be far more cautious about ratifying START II, let alone a follow-on agreement, than its Democratic predecessor. Nevertheless, during their September 1994 summit in Washington, Presidents Clinton and Yeltsin did commit the United States and Russia to early ratification of START II, to deactivate all weapons eliminated when the Treaty enters into force, and to explore 'the possibility, after ratification of START II, of further reductions of , and limitations on, remaining nuclear forces'.

Denuclearisation and Fissile-Material Controls
The United States and Russia made considerable progress in gaining greater control over fissile materials. A significant step was an agreement in March 1994 to conduct inspections at storage facilities for plutonium removed from nuclear warheads. It was also agreed that Russia would begin to phase out its three remaining reactors producing plutonium and prohibit the start-up of military reactors that have already ceased operations. Russia also agreed not to use any plutonium produced by those reactors still in operation for weapons. The agreement further called for verification arrangements to ensure compliance with these provisions, including on-site inspections of all military reactors that produced plutonium. These agreements brought Russia into line with the other four declared nuclear powers in ending the production of fissile materials for weapons purposes.

Cooperative efforts towards further denuclearisation were strengthened in September when, at the Clinton–Yeltsin Washington summit, the two countries agreed to exchange detailed information on aggregate nuclear-weapon stockpiles, on stocks of fissile materials, and on their safety and security. An agreement in principle on the details of this data exchange was reached in December 1994. Regarding nuclear weapons, the two countries agreed to exchange a detailed account of how many nuclear warheads each had produced since 1945, including separate tallies of each type of weapon. The data to be exchanged in 1995 and periodically updated thereafter would include the number of nuclear weapons already dismantled as well as those that will be dismantled as part of the arms-control agreements reached in the early 1990s. Data on the location of fissile-material storage sites will also be exchanged.

Although it has proven difficult to work out many of the details of these agreements, especially regarding verification arrangements, the agreed measures represent a concerted effort by Washington and Moscow to increase transparency of their past, present and future nuclear-weapon programmes. As a result, mutual confidence regarding both countries' nuclear intentions has been enhanced, as has effective control over nuclear weapons and materials. The latter is particularly important in view of an increasing number of reports during 1994 that illegal trafficking in nuclear materials is on the increase. The result has been enhanced cooperation between the two countries in countering this threat, as well as innovative measures designed to prevent smuggling at its source. For example, in an unusual operation launched in November 1994, the United States, with the assistance of Russia and Kazakhstan, secured a large stockpile of highly enriched uranium (sufficient to produce 25 weapons) by removing the fuel from Kazakhstan to the United States where it would be rendered unusable for nuclear weapons.

The ABM Treaty and Theatre Missile Defence

During the past year there was also much debate on the question of missile defences, an issue many had expected to disappear along with the Soviet Union. Growing fears of missile proliferation has brought the problem back in a new guise, and in tandem the question of what restrictions the ABM Treaty places on the development, testing and deployment of defence systems capable of intercepting missiles with less-than-strategic range has been raised. Although the ABM Treaty does not prohibit the testing or deployment of systems capable of intercepting tactical ballistic missiles, it does prohibit testing missile defence systems 'in an ABM mode'. The Treaty does not define what this means, but the US has long adhered to a set of guidelines that defined testing in an ABM mode as any test of a system against a target with an entry speed in excess of 2 kilometres per second (km/s) and flying at an altitude above 40km, which would be sufficient against tactical missiles of about 1,000km range.

In November 1993, the Clinton administration formally proposed that the United States and Russia should explore ways to clarify the ABM Treaty's provisions relating to non-strategic missile defences. The reason for this proposal was the US intention to develop a new system – the Theater High Altitude Area Defense (THAAD) system – capable of intercepting missiles with a range of 3,500km. Since development of THAAD would be prohibited under informal US guidelines, the administration sought to clarify the issue formally with Russia by proposing that a missile system be regarded as ABM-capable if it possessed a 'demonstrated' capability against a target entering with a speed in excess of 5km/s.

Since Russia is equally, if not more, concerned about the threat of missile proliferation, Moscow was inclined to accept clarification of the ABM Treaty in order to allow development of more capable anti-tactical missile defences. It therefore accepted the proposed testing limitation, but insisted that additional measures be added to ensure full and continuing compliance with the ABM Treaty's prohibition against deploying more than 100 ABM interceptors. Specifically, Moscow suggested in early 1994 that non-ABM interceptors be confined to a fly-out speed of 3km/s. The US responded in July with a new set of limitations: permitting deployment of ground- and sea-based interceptors with demonstrated speeds up to 3km/s and of air-based systems with speeds up to 5.5km/s; and permitting testing of sea-based interceptors with speeds up to 4.5km/s. Although Russia initially proposed modifying these limits, in November it suggested that the question of testing and deployment limits on interceptors with speeds in excess of 3km/s be deferred. It also proposed additional constraints on the number and location of missile systems that could be deployed and sought a formal prohibition on testing and deploying any space-based components.

The US–Russian stand-off was further complicated by the election of a Republican Congress in Washington in November 1994. The Republicans campaigned to revive the early deployment of a nationwide ABM system, and have since suggested that the entire US–Russian discussion on clarifying the ABM Treaty's provisions regarding non-ABM systems be suspended until the new Congress has fully explored the future of missile defences. Even if the Clinton administration can reach an accommodation with Moscow on these questions, the likelihood of Senate approval must be in doubt.

Non-Proliferation and Global Arms Control

Unlike the efforts between the United States and Russia, which demonstrated steady if limited progress, the promise of major advances in non-proliferation and global arms control that seemed apparent in recent years failed to materialise in 1994. With the notable exception of the breakthrough in US–North Korean negotiations, efforts to strengthen the nuclear non-proliferation regime through completion of a Comprehensive Test Ban Treaty (CTBT) were unsuccessful, calling into question the indefinite extension of the NPT in 1995. And while the first steps to strengthen the BWC were taken in 1994, the fate of the CWC remained uncertain as major countries (including Russia and the US) failed to ratify it.

North Korea

The one notable highlight in the nuclear non-proliferation picture of 1994 was the framework agreement between the United States and North Korea, which set the stage for resolving questions about Pyongyang's past behaviour and placing severe constraints on its future capabilities. Throughout the first half of 1994, North Korea and the international community were on a collision course. Pyongyang's refusal to allow routine and special inspections by the IAEA, continued uncertainty about the quantity of plutonium it might have produced surreptitiously, and mounting evidence that it was embarking on a major nuclear-weapon programme resulted in a concerted move by the United States to impose UN-mandated sanctions on Pyongyang in May 1994.

As so often in the past, however, North Korea stepped back from the brink, with former US President Jimmy Carter mediating a way out of the confrontation – a monitored freeze on North Korea's nuclear activities and high-level talks with the United States to resolve the nuclear issue. In October, after months of feverish negotiations – interrupted in July by the sudden death of Kim Il Sung, the North's self-proclaimed 'Great Leader' – a framework agreement was reached between the two countries which called for a phased resolution of the nuclear issue in return for normalisation of political and economic relations between the two countries. There will be four related steps:

- In return for a freeze and eventual dismantlement of North Korea's nuclear facilities (including three reactors and a reprocessing plant), the United States will arrange the supply of two more proliferation-resistant LWRs by 2003 and provide alternative energy sources (500,000 tons of heavy oil) until construction of the reactors is completed.

- The two countries will reduce trade and investment barriers, open liaison offices in Pyongyang and Washington, and eventually upgrade bilateral relations to ambassadorial level.

- The US will provide formal assurances to North Korea against the threat or use of nuclear weapons, while the North will resume negotiations with South Korea on their December 1991 bilateral denuclearisation agreement which, *inter alia*, calls for a ban on nuclear weapons and fissile-material production facilities to be verified through bilateral, on-site inspections.

- North Korea will remain a full party to the NPT and implement its safeguards agreement with the IAEA, including allowing on-site monitoring of the freeze on its nuclear facilities, *ad hoc* and routine inspections of facilities not subject to the freeze and, once a significant portion of the LWR project is completed but before key nuclear components have been delivered, accept any inspections deemed necessary by the IAEA to verify past plutonium production.

The framework agreement is an intricately crafted document that ensures that US concessions will only be made after North Korea implements key provisions. For example, the first shipment of heavy oil to the North took place only after the freeze on nuclear facilities had been implemented and monitored. Similarly, no significant nuclear components will be transferred to the North until the IAEA has completed any inspections it deems necessary to determine the precise amount of plutonium produced in the past, and placed whatever plutonium that has been produced under safeguards.

Nevertheless, the agreement came under heavy (and predictable) criticism from, among others, Republican members of Congress who suggested that the Clinton administration had made too many concessions in return for too little. As Congressional hearings made clear, however, the alternative to the deal was continued confrontation and the grim possibility of another war on the Korean peninsula. As a result, while criticism remained strong, there was no move to overturn the agreement or prevent the Clinton administration from implementing its part of the deal.

Comprehensive Test Ban Treaty

The year started with the general expectation that negotiations on a comprehensive test ban treaty in the Conference on Disarmament (CD) might be completed by the end of 1994 or early 1995. The United States, Russia, France and the UK all continued to abide by their declared testing moratoria (though China did not), and the Clinton administration had reversed 12 years of official US opposition to a CTBT. Reflecting this optimism, the CD voted in January 1994 to direct 'the Ad Hoc Committee to negotiate intensively a universal and multilaterally and effectively verifiable comprehensive nuclear test ban treaty'.

Almost immediately, however, negotiations in the CD stalled over differences on major issues, including the scope of obligations, the mechanism for the treaty's entry into force, and verification. The Conference failed to produce a draft treaty text when the 1994 session concluded in September, issuing instead a 100-page rolling text of heavily bracketed language on nearly all proposed treaty articles.

The most serious difference concerned the treaty's scope, that is, the actions to be prohibited. While most countries favoured a broad prohibition on 'nuclear-weapon explosions and nuclear explosions', China proposed an exception for peaceful nuclear explosions. Since peaceful nuclear explosions are often indistinguishable from nuclear-weapon explosions, allowing them would make the treaty unverifiable in practical terms. Another issue in contention related to a proposal to ban testing 'preparations' and computer simulations, because neither of these could be verified. There was also the question of whether the treaty would ban testing for all times. France and the UK proposed that it should allow safety and reliability tests 'in exceptional circumstances'. If adopted, this proposal would undermine the widely accepted aim of negotiating a non-discriminatory prohibition against nuclear testing. There was thus no support for this provision among the negotiating parties, and continued insistence on its inclusion would prevent rapid conclusion of the negotiations.

Verification has also been a contentious issue. There is disagreement about the number and type of monitoring systems that should be employed to verify compliance, with only on-site inspections and seismic monitoring acceptable to most parties. Concerns about cost and intrusiveness have prevented agreement on additional measures. And although it is agreed that verification should be managed by an international organization, differences remain over which organisation: the IAEA or some newly created body.

In view of these and other differences, the hope and expectation that a CTBT might be open for signature before April 1995, when the parties to the NPT were to convene to decide on the period of the NPT's extension, proved unjustified. In January 1995, the US finally launched a major initiative to complete negotiations as rapidly as possible. Washington dropped a controversial proposal, opposed by everyone else, that the treaty contain an 'easy-out' provision allowing any party to withdraw from the treaty after a specified period of time without having to invoke the 'supreme national interest' clause. The United States also extended its testing moratorium until 30 September 1996, and urged that negotiation continue without interruption at least until August 1995.

The NPT Review and Extension Conference

In April 1995, the more than 170 parties to the NPT were due to meet in New York to act on two basic determinants with regard to the Treaty, which came into force in 1970. Under Article 10 of the NPT, a majority of the parties must decide 25 years after the Treaty's entry into force (i.e., in 1995) whether it should be extended indefinitely or for a fixed period of time. All European and other advanced industrialised countries favour the indefinite and unconditional extension of the Treaty, while many other countries remained undeclared on their position until the eve of the conference. By the end of March, it appeared likely that a slim majority of parties backed the indefinite extension of the Treaty and supporters of this position believed that once a majority had been declared, other states would join the growing consensus.

Whatever the eventual outcome of the NPT review and extension conference, the deliberations were sure to be dominated by two issues: first, the degree to which parties were upholding their commitment under Article 4 to promote the peaceful use of nuclear energy; and, second, the extent to which the nuclear powers had fulfilled their obligations under Article 6 to halt the arms race and move towards nuclear disarmament. Iran has made itself the champion of the first of these issues. It argues that its full and verified compliance with the NPT makes illegal the US effort to prevent the export of nuclear technology to Tehran. In the preparatory meetings for the extension conference, Iran has skilfully played on this concern and gained a sympathetic hearing from many other non-aligned countries. But its efforts are unlikely to derail prospects for an indefinite extension of the Treaty.

The second issue has long dominated NPT review conferences, and this year is unlikely to be an exception. The major difference compared to previous years is that the end of the Cold War has reduced the salience of the traditional complaint that the nuclear powers were not fulfilling their Article 6 obligations. The entry into force of START I and anticipated ratification of START II moved the two largest nuclear-weapon states along a path of sharp nuclear reductions, while the ongoing negotiations on a CTBT and proposed negotiations on ending fissile-material production for weapons purposes suggested a commitment by the nuclear powers to cap their ability to produce and deploy additional nuclear weapons.

Though significant, these steps have not yet satisfied some non-nuclear states, who point out that post-START US and Russian force levels will still exceed those deployed at the time the NPT entered into force. In addition, it is argued that the other declared nuclear powers have yet to join the disarmament process, that the CTBT remains to be completed, that binding security assurances have yet to be negotiated, and that nuclear weapons continue to fulfil a central role in the security policies of the nuclear states.

In the end, however, a changing world and continuing progress on arms control are likely to undercut these arguments. The most powerful reason for the indefinite extension, and one the US and other major countries are likely to stress, is that the NPT enhances the security of non-nuclear states as much as of the nuclear states by reducing uncertainty about the potential nuclear capabilities and intentions of neighbouring countries. It is this argument, ultimately, that will convince most NPT parties to support its indefinite extension.

Chemical Weapons Convention

The CWC was opened for signature with much fanfare on 13 January 1993, with 130 countries attending the signing ceremony. Under the terms of the convention, the CWC would enter into force 180 days after 65 countries had ratified it or two years from the date of signature, whichever is later. Since that time, 159 countries have signed the convention, but only 19 (including Australia, Bulgaria, Germany, Greece, Mexico, Norway, Spain and Sweden) have completed ratification procedures. As a result, the convention did not enter into force at the earliest possible time and must now await ratification by over 40 other countries.

An important reason why a sufficient number of countries did not ratify in 1994 was the failure of the United States and Russia (states with the largest known stockpiles of chemical weapons) to complete their ratification procedures. It is generally assumed that once these two countries, and especially the United States, ratify the convention, others will follow suit. Indeed, some countries like France have completed ratification but are awaiting similar US action before depositing their instruments of ratification. These countries want to avoid a repetition of the fate of the 1925 Geneva Protocol, which banned the use of chemical and bacteriological weapons. Then, the US was also actively engaged in the negotiations, but it failed to ratify the protocol for 50 years.

The United States had every intention of ratifying the CWC in 1994 – indeed, by July – to ensure the earliest possible entry into force of the convention. President Clinton repeatedly called on the US Senate to complete its ratification procedures, and hearings before the Senate Foreign Relations, Armed Services and Intelligence Committees were completed by early July 1994. But ratification hung on a number of difficulties, with some bearing little relationship to the subject of the convention. Some Senate committees were slow to act once the hearings had been completed, with two (Foreign Relations and Armed Services) even failing to complete a final report before Congress recessed for the year in October 1994. Partisan politics were also to blame. In the final months before the mid-term elections, the Republicans were loath to grant the President a political victory, and a final vote on the CWC, as on so much else, was held up. Some Senators seized on US concerns relating to Russian chemical-weapon activities to delay full consideration of the convention. Concern focused in particular on incomplete data regarding past and present chemical weapons capabilities and activities that Russia provided to the United States as required under the second phase of the 1989 Wyoming memorandum of understanding on chemical weapons.

The Russian failure to ratify the CWC was also noteworthy. The main concern in parliamentary deliberations on the convention related to the cost of destroying the vast stocks of chemical weapons and agents Russia inherited from the Soviet Union. The cost is expected to be in billions of US dollars, and there is currently no plan to finance this requirement, which must be implemented no later than 10–15 years after the convention's entry into force. Although the US has devoted considerable resources to help Russia develop safe and environmentally sound destruction technology, the bulk of the cost will have to be borne by Russia.

In the end, the future of the CWC will rest on the US completing its ratification procedures. There is every reason to believe that a sufficient number of votes in favour of the convention can be garnered within the Senate, even one controlled by the Republicans. Senator Jesse Helms, the new Chairman of the Foreign Relations Committee and a long-time opponent of arms control, has turned to Senator Richard Lugar to steward upcoming arms-control treaties, including the CWC, through the Senate. Senator Lugar is on record in support of the convention. But the Senate will have to act, sooner rather than later, if the CWC is to be spared the fate of the Geneva Protocol. Once it does, however, there is every reason to believe that other countries will follow rapidly. This will likely include Russia, whose power and influence over the future direction and implementation of the convention (including critical issues relating to cost and inspections) depends on it being an original party to the convention, which requires ratification before the CWC enters into force.

Biological Weapons Convention

In September 1994, a Special Conference of the states parties to the BWC convened to consider a report on potential verification measures for the convention prepared by a group of governmental experts in the previous two years. Two main issues dominated the conference. First, a number of developing countries (notably China, India and Iran) sought to use the occasion to underscore their long-standing position that export controls by the industrialised countries were in contravention of Article 10 of the BWC, which provides for the free flow of information and technology for peaceful purposes. The second, and more substantive, issue concerned the form and possible content of any verification measures that might be negotiated. With the United States having abandoned its previous opposition to negotiating measures designed to enhance transparency of biological warfare activities and strengthen compliance with the BWC, the main issue concerned modalities of negotiation, timing and the content of the measures. In the end, it was decided to establish an Ad Hoc Group open to all parties, which would consider 'appropriate measures, including possible verification measures' and draft specific proposals designed to strengthen the convention in a legally binding manner. The Special Conference failed to establish a specific timetable, however, noting merely that work would be completed as soon as possible and a report submitted to the fourth BWC review conference in 1996 or later at another special conference.

1995: A Critical Year

The limited, though important, advances during 1994 have kept the arms-control train rolling. But 1995 is likely to be the year in which it is decided whether the train continues on its steady course made possible by the end of the Cold War or derails instead, thereby calling into question the progress achieved to date. In 1995, the United States and Russia face the choice between ratifying the START II Treaty and beginning a new round of talks aimed at still further nuclear reductions on the one hand, and reversing, or at least halting, the nuclear build-down both countries have so far endorsed on the other. In part, this choice will depend on the fate of the ABM Treaty, which is once again being called into question in parts of Washington and on whose maintenance further nuclear reductions will in part depend.

It will also be a critical year on the proliferation front. The Nuclear Non-Proliferation Treaty is up for review and its duration must be extended. Indefinite and unconditional extension will give all states a greater degree of confidence that nuclear weapons will not spread inexorably. Uncertainty about the Treaty's fate, on the other hand, will provide many states with an incentive to hedge their bets, creating new sources of distrust and possibly new reasons to maintain or acquire the capability and expertise rapidly to build a nuclear arsenal. Moreover, the scourge of chemical weapons may finally be eliminated in 1995 if a sufficient number of states ratify the convention. But as time moves on and key countries hesitate, faith in the CWC will dissipate, weakening efforts to eliminate these weapons from the face of the earth. It has been said that with the end of the Cold War, arms control no longer has a future. It is up to the major powers in the world to prove that such prognostications are wrong.

The Americas

By choice, US President Bill Clinton was not an active participant in foreign policy and most of the stands he did take were unsuccessful. A signal exemption was negotiating and assuring passage of the North American Free Trade Agreement (NAFTA). Although NAFTA is not intended to provide even the basic framework for any political integration, its creation will bring about a new sense of regionalism in the Americas. The landslide victory of the Republican Party in the mid-term elections in the United States has brought into being a stronger constituency opposed to even the modest regionalism that NAFTA represents. Their fears were enhanced at the beginning of 1995 by the negative economic developments in Mexico.

Latin American societies in general are societies in transition from mostly authoritarian military rule with controlled economies to democratic political governments championing free-market economics. In each of them there are serious strains of both an economic and political nature. It is encouraging that under the circumstances none have returned to the dark practices of the recent past. But most are struggling. The strains were set out in sharp relief in Mexico, where assassination of its leaders, rebellions by its poorer citizens and collapse of the value of its currency dwarfed the problems in the rest of the continent. Help from abroad, particularly from the United States, has brought temporary improvement. As is the case in the rest of Latin America, however, it appears that the short-term problems will be surmounted, if with difficulty, and the longer term promises much improvement.

The US: The Unilateralist Temptation

Conventional wisdom had it that President Bill Clinton was a savvy domestic-policy expert with a strong mandate for change, but a foreign policy novice with little interest or ability in global affairs. By the end of 1994, however, there was little evidence that these simple clichés were at all accurate. During 1994, the Clinton administration's ambitious domestic reform agenda was rejected by the American people and Congress, while in foreign affairs the administration achieved some surprising successes. Whatever the relative merits of the President's actions at home or abroad, US policy took a dramatic turn in November 1994 with the election of a

Republican-dominated Congress opposed to the President's policies on both fronts. The US began 1995 with a Congress determined rapidly to impose a populist and conservative economic and social agenda and to reduce America's multilateral commitments abroad, all against the administration's wishes. After an interlude of just two years, divided government had returned to Washington.

Clinton's Domestic Travails

Reform of America's costly and incomplete health-care system was Clinton's highest domestic priority for 1994. He hoped to guarantee coverage for all Americans while at the same time making significant savings by streamlining the system, cutting down on doctors' paperwork, making unnecessary the use of costly emergency services by the uninsured, and requiring employers to pay a significant share of their employees' insurance costs. But Clinton's far-reaching goals proved impossible to achieve. Perceiving universal care as too costly, and employer mandates as a threat to small businesses, Republicans (and many Democrats) in Congress rose in concerted opposition to the plan, proposing in its stead a range of compromise options that Clinton would not accept.

The President eventually backed away from his insistence that immediate 100% coverage was central to health reform. In jettisoning this key element he sought support for alternatives, but by then it was too late. Congressional and ideological opponents, aided by consistently negative television commercials sponsored by the insurance industry, managed to portray the Clinton plan as exceedingly bureaucratic, expensive and anti-choice. The American people decided they preferred the devil they knew to one they did not. Clinton's health-care plan not only failed to pass in 1994, it did not even come to Congress for a vote.

The President had more success with his legislation on crime. In August 1994, the first crime bill in six years was passed, banning several types of assault rifle, providing nearly $30 billion for additional police recruitment and new prisons, expanding the use of the death penalty, and guaranteeing life sentences for anyone convicted of three federal crimes ('three strikes and you're out'). Success came after fierce administration lobbying, and was a high point in an otherwise dismal legislative year. Welfare reform, Clinton's third social priority, suffered the same fate as health care: it never made it to the House or the Senate floor.

To compound the President's problems, scandals and attacks on his character continued to plague the White House. In May, Paula Jones, an Arkansas state employee, filed a public lawsuit alleging that she had been sexually harassed by then-Governor Clinton. Whatever the validity of the allegation it reinforced Clinton's image as a philanderer. Clinton's reputation was also damaged by the 'Whitewater affair', a complicated tale of a failed Savings and Loan and a misguided investment that raised questions about his possible abuse of political power for personal gain while governor of Arkansas.

During that same period Hillary Clinton apparently made $100,000 betting on cattle futures helped by traders who needed her husband's political support, a fact that did not help the first couple's reputation. This was further tarnished when Congressional hearings on the Whitewater affair were held in July. A number of senior administration officials were publicly grilled, and several were forced to resign.

By autumn 1994 the administration's domestic agenda was in disarray and the President had lost the public's support. In the November mid-term elections, Republicans gained a 53–47 majority in the Senate and a 230–204 majority in the House, giving them a majority in both Houses of Congress for the first time in 40 years. Republicans also made a net gain of 11 governorships, giving them control of the eight largest US states (except Florida). Unlike the 1992 election, this was not simply a landslide against incumbents but a landslide against Democrats. All the Congressional Republican incumbents won their seats while the rate of re-election for Democrats was just 85%, well below the normal rate. This figure looked more respectable than it actually was, however, since many sitting Democrats chose not to run at all for fear of defeat. Even Tom Foley, the respected Speaker of the House, was defeated, and had to give his powerful position to Newt Gingrich, a firebrand populist who pledged to change radically the way Congress worked. Another unique aspect of the Congressional election was its national (rather than local) character: Republican candidates to the House ran on a common 'Contract with America' that pledged the party to a programme of Congressional term limits, sharp cuts in welfare programmes, a balanced budget amendment to the Constitution, and a tax cut for the middle class.

Why did Clinton's domestic agenda fail? Perhaps Americans had trouble accepting the generational change that this young southern president represented. The first post-Second World War president not to have served in the armed forces, he had, like many of his peers, opposed the war in Vietnam. Moreover, his reputation for being incautious and impetuous, aggravated by the Whitewater and Paula Jones affairs, continued to claim the public limelight and hampered the development of public confidence in his presidency. Clinton's style of leadership remained disorganised and free-form even after two years in the White House, adding to the impression of indecision and ineffectiveness.

Most importantly, Clinton also overstretched the mandate he was given in 1992 when he was voted in by just over one-quarter of those of voting age. Although he and his advisers recognised the limitations of that mandate, they nevertheless attempted to force far-reaching change on an electorate that may have been ready for action, but not for a return to the liberal, interventionist and anti-middle-class agenda that Clinton's priorities seemed to represent. For the first time in recent memory a president's popularity fell while the economy performed well. Not even a 4% growth rate, the lowest inflation for decades, and falling unemployment could save the administration from defeat, or the country from divided government again.

Foreign Policy: the Retreat from Multilateralism

Clinton's foreign-policy record stood in contrast to his performance at home. After a troubled first year and a half marked by clashes between the President and the military, tragic failures in Somalia and Yugoslavia, and some high-profile waffling, turnarounds and internal divisions, there followed in the second half of 1994 a string of foreign-policy successes and a more coherent theme: 'marrying diplomacy with force'.

The invasion of Haiti that had been so widely criticised in Congress and by the media turned out to be a spectacular short-term success, with no American casualties and the restoration to the presidency of the illegally deposed Jean-Bertrand Aristide. In the Persian Gulf, Clinton responded with determination to a renewed attempt by Saddam Hussein in October to threaten Kuwait, and Iraq backed down. The administration also secured a deal to reverse North Korea's suspected nuclear-weapons programme; saw through the implementation of its denuclearisation policy in the former Soviet states with the entry into force of the Strategic Arms Reduction Talks (START I) Treaty and Ukraine's accession to the Nuclear Non-Proliferation Treaty (NPT); continued its good record in foreign economic policy by overcoming Congressional resistance to the ratification of a General Agreement on Tariffs and Trade (GATT) initiative and by expanding the role of Asia-Pacific Economic Cooperation (APEC); kept the fragile Middle East peace negotiations on course; and even seemed to play a role in advancing peace in Northern Ireland.

While these developments appeared to herald a new coherence and competence in the administration, the Secretary of State, said to be quietly effective on certain issues, continued to be ineffective in articulating a long-term vision of America's role in the world. And the President remained preoccupied primarily with domestic and political issues.

Into this vacuum stepped former President Jimmy Carter, whose one-man travelling show to North Korea, Haiti and the former Yugoslavia bore some fruit, but whose actions raised worrisome questions about who was in charge of US diplomacy. Carter was potentially useful in testing out a possible solution to these conflicts, a solution that the administration could theoretically choose to accept or disavow, but his independent spirit and fundamental belief in ultimate human goodness – even of people like Kim

Il Sung, Raoul Cedras and Radovan Karadzic – made it unclear whether, once turned loose, he could really be controlled.

The sustainability of Clinton's successes, in any case, was far from guaranteed, and the administration's continued tendency to delay taking decisions indicated a propensity to put off problems rather than solve them. Clinton's foreign-policy record may no longer be the political liability it was during 1993, but to count on it to boost his popularity while domestic gridlock prevails would seem a highly tenuous political strategy for the coming two years.

Backing Away from the UN

More significant than the relative success or failure of American foreign policy in 1994 was its changing character – during the year the Clinton administration departed rapidly from the platforms of multilateralism and humanitarianism that it had initially pursued. The country's bad experience in Somalia (the bulk of American troops left, mission unaccomplished, in March 1994) was reinforced by the continued failure of United Nations (UN) peacekeeping in the former Yugoslavia and in Rwanda.

Americans began to question the early Clinton vision of a world in which US power would be deployed multilaterally, in the name of human rights, democracy and international law. Indeed, these events gave many Americans the impression that the US succeeded when it acted alone or as a leader – as in Haiti, Iraq and North Korea – and failed when – as in Bosnia or Somalia – it tried to work as part of a UN force. Whether this axiom is accurate or not is irrelevant; the impression may dominate American foreign policy for years to come.

In May 1994, Clinton issued Presidential Decision Directive 25 (PDD-25), a comprehensive attempt to set out – but also to limit – the circumstances under which the United States would support a UN operation or commit its own troops abroad. The document stated that the United States would only support UN peacekeeping missions that served US interests, were backed by adequate means, and were accompanied by realistic criteria for coming to an end. PDD-25 also made it clear that the US would only commit its own troops to action if several preconditions were met: that the risks to American personnel were deemed acceptable; that personnel, funds and other resources were available; that US participation was necessary for success; that Congressional and public support could be marshalled; that the mission had clear objectives; and that command-and-control arrangements were acceptable. Aside from the ambiguous nature of many of these criteria (who defines what is 'acceptable' or 'available'?), it was difficult to envisage many cases that would meet all the requirements.

The tendency to back away from participation in UN operations was only reinforced by the continued US experience in the former Yugoslavia. Since coming to power in January 1993, the administration had advocated, with varying degrees of consistency, lifting the UN arms embargo on the Bosnian government, and the more active engagement of North Atlantic Treaty Organisation (NATO) air power against the Bosnian Serbs. Unlike the Europeans who were taking the lead in UN peacekeeping in the former Yugoslavia and who saw the conflict largely as a civil war (albeit with one side – the Serbs – as most responsible), many Americans believed the war to be one of unilateral aggression, which demanded a more vigorous military response. The Americans were unwilling to engage their own troops on the ground, however, and could offer little more than periodic and limited air-strikes (such as those threatened around Sarajevo in February 1994 and those made near Gorazde in April), which the Europeans usually resisted for fear of compromising their avowed neutrality as UN forces and hence suffering possible retaliation against their troops on the ground.

Although the US was able to broker a federation between the Bosnian Muslims and the Croats, American frustration with the European approach built up as the year went on and the carnage continued. Congress tired of administration arguments that it could only act with the agreement of its European allies or the UN, and decided to act on its own. On 11 May, the Senate passed a resolution calling for a unilateral lifting of the arms embargo against the Bosnian government, prompting President Clinton to respond that this 'would kill the peace process, sour our relations with our European allies in NATO and the UN, and undermine the partnership with Russia'. On 10 August, however, Clinton announced that if the Bosnian Serbs did not accept the peace plan put forward by the five-nation 'Contact Group' by 15 October, the United States would ask the UN Security Council (UNSC) to lift the arms embargo. At the same time, he promised Congress that if the Serbs failed to agree by 15 November, the United States would act alone.

This fateful step took place on 11 November. The US outraged its European allies when it announced that it would no longer participate in *Operation Sharp Guard*, the NATO mission preventing arms from reaching the former Yugoslavia, and that no further US intelligence on the arms embargo would be shared with the Europeans. Aside from raising the absurd possibility that the US might now be willing to break an embargo that NATO was enforcing – would British or French forces fire on American ships? – the action sent a warning to Europe that for Americans, the importance of the Alliance was fading.

The Clinton administration did its best to minimise the significance of the incident, and, apparently concluding that NATO was more important than Bosnia, by December had backed away from its own efforts to lift the embargo or undertake air strikes against the Serbs. Congress, though, was less conciliatory: Senator Bob Dole said on 11 November that 'the UN should get off NATO's back', and that NATO had become 'almost irrelevant' because of its inability to challenge the Bosnian Serbs. The former Yugoslavia had tarnished the reputation of multilateralism and was eroding American support for NATO and the UN alike.

Interventionism, if not multilateralism, fared better in Haiti. The Haiti problem, which had been a thorn in the side of the administration throughout its first year in office, became increasingly intolerable during 1994. The US-backed strategy of sanctions crippled the island's economy and increased its already appalling poverty, but was unable to achieve its primary aim – to oust the military regime led by General Raoul Cedras that had come to power in a coup against President Aristide in September 1991. By mid-summer, with the administration's embarrassment increasing as it decided forcibly to repatriate the flood of Haitian refugees created largely by its own sanctions (a policy that Clinton had labelled as 'criminal' when conducted by George Bush), Clinton began to hint that he would use military force if the Haitian junta did not leave power voluntarily. This eventuality was backed by the UNSC on 31 July. But after the retreats in Somalia and the empty threats in Bosnia, US warnings lacked credibility, and Cedras did not budge.

Despite overwhelming Congressional and public opposition to an invasion of Haiti, the administration felt it could not back down. On 19 July, President Clinton announced that the Cedras regime's 'time was up'. A last-minute negotiating effort led by former President Carter and including Senator Sam Nunn and former Chairman of the Joint Chiefs of Staff Colin Powell seemed on the brink of failure when, with the 82nd Airborne Division already in the air, Cedras finally agreed to leave. Pundits and Republicans in Congress bemoaned the dangers of even an unopposed invasion, and others complained that Carter had caved in by agreeing to cooperate with the Haitian police and let Cedras stay in the country. But as time went on the mission's success became clear. On 13 October Cedras formally resigned and left the country, and within a month Aristide was back in power. At the end of March 1995, the American troops, having lost only one soldier in hostilities, handed the operation over to a primarily UN force as planned.

Of all the areas of US foreign policy, Haiti and the Balkans received the most public attention. In other areas, quiet progress or simple continuity prevailed. The administration clung to its policy of 'dual containment' in the Persian Gulf, resisting pressure from France and Russia to convince the UN to lift sanctions against Iraq, while insisting at the same time on tougher international measures to ostracise Iran. Many saw dual containment as a 'non-policy', but few had anything more constructive to offer in its place. Continuity also prevailed in policy towards Russia, where the administration did not waver in its support for reform, despite concerns that the Yeltsin administration was headed in an anti-democratic direction after Russian troops invaded Chechnya in December 1994. The Chechen affair served to energise the transatlantic debate over European security and NATO expansion that Clinton had advanced with the launch of the Partnership for Peace (PFP) initiative in January 1994. By late that year he saw fit dramatically to accelerate planning for NATO expansion, causing some consternation among European allies who thought they 'had a deal' with PFP. Clinton also kept to his course of seeking to promote regional economic cooperation and free trade in Asia, and he scored some successes, including a financial services agreement with Japan in January 1995 and a pledge at APEC's Jakarta summit in mid-November 1994 to create an Asian free-trade zone by 2020. The primacy of economics also led the administration to back down from its threat to remove, because of human-rights violations, China's most-favoured nation status, which was renewed on 26 May 1994.

The Republicans in Power

A crucial factor in the further development of Clinton's foreign policy was the election of a Republican Congress determined to impose its own, and very different, agenda on the President. The Republicans had nothing against 'marrying diplomacy with force', but they differed with the administration over the situations in which force should be used – to the Republicans, spreading democracy in the Third World and peacekeeping on behalf of the UN were not worth risking American lives.

Many in the new Congress opposed the nuclear deal with North Korea, threatened to withdraw support for the Haiti operation, accused President Clinton of weakness in dealing with Russia, called for unilateral US actions in Bosnia, demanded more spending on defence and less on foreign aid, and rejected the administration's (albeit fading) inclination to work closely with the United Nations. The new majority called for the rapid expansion of NATO to include Central Europe, an enhanced ballistic-missile defence programme, and a ban on US troops being placed under UN command. Clinton would now have not only to tackle a still-difficult foreign-policy agenda, but to do so knowing that the purse strings for international affairs were in the opposition's hands.

Defence and Non-Proliferation Policy

The centrepiece of the Clinton administration's defence policy, the 'Bottom-Up Review', came under pressure during 1994. The Review, announced in September 1993, laid out the force-structure and readiness levels necessary to win two major regional contingencies while at the same time maintaining resources for lesser conflicts and investing in future capabilities. In February 1994, the administration put forward a $263.7bn defence budget that it claimed would enable the US to fulfil its military commitments. These included funding for large procurement programmes such as the F-22 and F/A-18E/F fighters, the *Arleigh Burke* class of destroyer, the CVN-76 nuclear-powered aircraft carrier, and the V-22 *Osprey*, a tilt-rotor aircraft for the Marines.

Congress, unsurprisingly, had pet funding projects of its own. It was concerned that the administration's budget was deficient primarily in the areas of strategic lift and bomber capabilities. Thus, after cutting out some of the President's requests, it added several hundred million dollars for the Sealift Fund, approved production of the problem-plagued C-17 airlifter, and loaded on $125 million to preserve the option of ordering more B-2 stealth bombers than the 20 sought by the administration. The budget passed with little fanfare in September, with the total allocation nearly unchanged at $263.8bn.

The defence spending debate, however, was not over. Republicans and other critics claimed that 'operations other than war', such as the invasion of Haiti or the Yugoslav no-fly zone, were eating into funding for current readiness and undermining future preparedness – creating the 'hollow' army from which the United States suffered in the 1970s. As evidence of the problem, and seeking to score political points, the critics pointed to the inadvertent April 1994 downing by American F-15s of US helicopters over Iraq and the low readiness rating of three of the army's 12 divisions.

It was true that the operational tempo of US forces – the time soldiers spend on training or military operations away from their home base – was higher than it had been for decades, and was using up resources intended for other purposes. The administration, however, pointed out that spending on contingencies other than the major regional ones during 1994 (including the former Yugoslavia and Haiti) was less than $2bn out of a total budget of $261bn, and that the administration had fully funded all service requests for operations and maintenance spending, which at $88bn dwarfed the cost of the peacekeeping operations.

Whatever the current state of US military readiness, questions about future capabilities – in the wake of reductions in personnel, procurement, and research and development – remain. Indeed, the General Accounting Office warned in July 1994 that higher-than-expected inflation, unforeseen costs of environmental clean-up at closing bases, the high operational tempo for US forces, and a Congressionally mandated pay rise for military personnel would result in a $150bn shortage in the $1,295bn projected budget in the Pentagon's Future Years Defense Programme until 1999. The

Pentagon insisted that the shortage was only $40bn, and in August proposed delays in the procurement of some of its major weapons programmes to compensate.

Sensing the political importance of the issue (and certain that the new Congress would seek to increase defence spending anyway), Clinton in December gathered the Joint Chiefs of Staff on the White House lawn to announce a proposed $25bn increase in defence spending over six years (1995–2001), largely to cover the costs of pay rises and programmes to improve the quality of life for soldiers. The President also announced that the administration would seek a supplemental appropriation of more than $2bn for Fiscal Year 1995 to replace funds spent on contingency operations such as Haiti. Later in September, Defense Secretary William Perry announced cuts or delays to weapons modernisation programmes designed to yield a saving of $7.7bn over the 1995–2001 period. The shift of funds from procurement to operations will put a crimp in force structure modernisation plans.

Curbing the proliferation of weapons of mass destruction – nuclear, biological and chemical as well as their delivery systems – was one of the United States' top priorities of the year. In addition to reaching the major denuclearisation agreements with North Korea and the countries of the former Soviet Union noted above, the administration budgeted $400m to assist the former Soviet states with the dismantling of their nuclear and chemical weapons (the Nunn–Lugar programme), and an additional $33.5m for its Defense Counter-Proliferation Initiative, the Defense Department's programme for confronting threats from weapons of mass destruction if non-proliferation should fail.

In September, the administration published its long-awaited 'Nuclear Posture Review', commissioned by Clinton's first Defense Secretary, Les Aspin. This Review concluded that it would be unwise for the United States to drop below the START II level of 3,500 nuclear warheads until Russia's future was more certain. While the Review did recommend some nuclear reductions – leaving the country with approximately 14 *Trident* SSBNs, 86 intercontinental bombers (66 B-52s and 20 dual-capable B-2s) and 500 *Minuteman* III single-warhead intercontinental ballistic missiles – it made no significant changes in nuclear declaratory policy, such as adopting a 'no-first use' pledge. Critics felt the Review was too conservative and that it failed to seize the opportunity of moving towards a non-nuclear world (especially with the major NPT review conference scheduled for spring 1995), but the administration was too deeply focused on non-nuclear issues to take any decisive new steps towards strategic reductions. The Republican Congress is likely to be even less prepared to move towards further radical reductions.

The administration was also active – and had more Republican support – in the area of theatre missile defence. Committed to ensuring the protection of US troops who might be deployed within reach of enemy short- or medium-range ballistic missiles, it gave top priority to developing its Theater High-altitude Area Defense (THAAD) programme as well as a number of other land-based, air-based and sea-based interceptor systems. At the same time, the administration has sought an agreement with Russia on interpretation of the 1972 Anti-Ballistic Missile (ABM) Treaty that would allow for the deployment of such theatre systems without threatening the ban on strategic missile defences, which the Americans profess still to support.

Critics of the theatre missile defence programme claim that the administration's support for THAAD and similar programmes threatens to undermine both the ABM Treaty and the START agreements, and that the administration will have to choose between two conflicting goals (arms control and missile defence). Secretary of Defense Perry denies that such a choice exists, but has hinted that missile defence would not be sacrificed, even if a choice had to be made.

America's Temptations

As the November 1994 mid-term elections showed, Americans are deeply dissatisfied with the state of their country both at home and abroad. They are frustrated by the complexity of conflict in the post-Cold War world (and the thanklessness of dealing with it), and have become disillusioned with the UN, disappointed by allies, and unwilling to shoulder the burdens of a lone superpower role. Many observers fear that the reaction to this frustration will come in the form of isolationism, a refusal to take responsibility for international security and a determination to act in the short-term good of the United States alone.

Just as likely as nostalgia for an era when Americans did not deal with the rest of the world, however, is nostalgia for an era when they largely ran it. The American temptation today is unilateralism as much as isolationism: international engagement on American terms and in American interests, or no international engagement at all. The unpleasant reality is that neither course is good for the rest of the world, nor for the United States itself.

Troubled Governments in Latin America

For the democratic governments of Central and Latin America economic liberalisation, primarily through privatisations, financial market reform and regional trade integration, remained a top priority. They struggled, however, with financial crashes and scattered political opposition, and often moved ahead only at the cost of subverting the institutional formalities of legislative and judicial review. New monetary problems swept the region, with the World Bank classifying 10 of the 28 Latin American and Caribbean countries as severely indebted. This economic hole is likely to deepen because the Mexican monetary crisis will decrease inflows of portfolio capital. In the face of these economic difficulties, eight Latin American countries held open and contested presidential elections in 1994; in all of them, the more conservative candidates won through.

Murder and Money Problems in Mexico

Despite efforts at reform, Mexico struggled with a series of threats to its economic and political stability in 1994 and the early part of 1995. In 1994 it opened its financial markets and granted at least 18 foreign banks, 16 stockbrokers, 12 insurance companies and one leasing group authorisation to operate in Mexico. The pace of state-owned company privatisation continued, with telecommunications, airports, railways and petrochemicals slated for sale. Mexico's gross domestic product (GDP) growth reached a respectable 3.3% in 1994. Yet, despite increased interest rates of 60%, its current-account deficit of $25bn and its 7% inflation promised to worsen in 1995.

The new administration of Ernesto Zedillo Ponce de Leon, inaugurated in December 1994, immediately faced a monetary crisis when Mexico's policy of maintaining an overvalued peso, opening financial markets to short-term foreign capital, and relying on these portfolio investments to cover high budget deficits backfired. The Mexican peso was devalued and allowed to be freely set, but it quickly plunged more than 30% against the dollar. Hard currency reserves were depleted in an abortive attempt to support the peso, leaving the government unable to fund the short-term, fixed-income dollar securities – *tesebonos* – which matured in early 1995. The government was unable to formulate a policy to reassure investors of its ability to pay these cheques, and thereby stabilise the peso and halt the flight of foreign capital.

The US government and the International Monetary Fund (IMF) put together a $50bn international financial rescue package in late February 1995. It required, among other things, effective US control of Mexico's oil export earnings, but it failed to calm financial markets. The risk of Mexican company and bank failures remained high despite a January 1995 wage renegotiation between the Zedillo administration and the labour unions.

The economic and financial difficulties of the end of 1994 were preceded by political turbulence. The presidential electoral campaign was rocked in March 1994 by the assassination of the governing Institutional Revolutionary Party (PRI) candidate, Luis Donaldo Colosio. There were suspicions that the murder was inspired by infighting within the PRI, where anti-reformers were mounting a rearguard action against the principles for which Colosio stood. With the loss of his favourite to succeed him, President Carlos Salinas de Gortari turned to Ernesto Zedillo, a fairly colourless technocrat. The October presidential elections went ahead as scheduled and the PRI candidate won by a little more than 50% of the vote.

The much lower than usual vote for the PRI reflected the political changes under way in Mexico. The post-election killing of the PRI's Secretary-General, José Francisco Ruiz Massieu, further destabilised the economy, leading to an $11bn outflow of capital even prior to the December peso devaluation. Investigators have linked this assassination to charges of corruption and drug trafficking levelled at Massieu's brother, former Deputy Attorney General Mario Ruiz Massieu, as well as former President Salinas' brother Raul Salinas de Gortari. Zedillo's eforts to distance himself from the charges of corruption and poor economic management of the previous administration created a public rift between him and Salinas, leading Zedillo to force the former president into indefinite exile.

These developments were played out against the backdrop of continued political violence and unrest, which further undermined confidence in the Mexican economy. Just as NAFTA was scheduled to come into effect, the government found itself forced to deal with an uprising in January 1994 by the guerrilla organisation *Ejército Zapatista de Liberación Nacional (EZLN)*, composed primarily of native peasants of Mayan descent, in the Chiapas region. The Zapatista rebellion, originally launched to demand democratic reforms and increased rights for indigenous people, has been joined by a number of other groups that share their goals of land redistribution and electoral reform.

The guerrillas were driven from six towns by 15,000 troops hurried to the region, but it became clear that a military solution was not in sight when they melted into the hills and vowed to continue the struggle. President Salinas moved quickly to meet some of the rebels' political demands. The leaders of nine political parties joined the ruling party in an agreement on sweeping electoral reforms. Salinas also appointed a left-of-centre PRI leader to negotiate with the rebels. His efforts, combined with promises of economic and social change in the rural areas, put out the fire temporarily, but discontent smouldered throughout the rest of the year.

In early February 1995 the government sent thousands of troops back into Chiapas in an attempt to capture the leader of the rebels, self-styled *subcomandante* Marcos, and to crush the rebellion. Although the operation did capture a number of prominent members of the *EZLN* and recovered much of the rebel-held territory, it was not a complete success. Not only was Marcos not captured, but the Mexican Army could not stop him from broadcasting his message to the world over the InterNet as he has for the past year. And the guerrillas, who had sufficient warning of the troop movements, rather than fight hopelessly against an overwhelming force, faded again into the hills perhaps to fight another day.

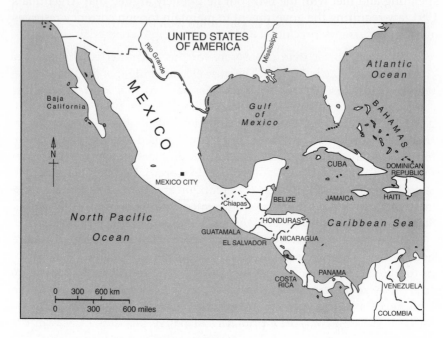

Like his predecessor, President Zedillo turned to the political arena to counterbalance these pressures. The Mexican government and the main opposition parties agreed to implement new democratic reforms, including controls on political fund-raising and campaign spending and to ensure the independence of the federal election agency. In addition, there was agreement on holding new state elections in the southern states of Chiapas and Tabasco where leftist candidates asserted widespread fraud in the October 1994 elections. The PRI governor of Chiapas was forced to step down and the opposition National Action Party (PAN) won an important governorship in the neighbouring state of Jalisco. With the government troops ensconced in most of the state's villages, the situation has calmed considerably in the Chiapas area, but the potential for renewed hostilities in the southern part of the country remains high.

Unsettling Argentina and Brazil

In the aftermath of the January 1995 Mexican monetary crisis, investors feared that the Argentine and Brazilian economies would suffer similar fates, and the stock-markets in both countries slumped dramatically. Confidence in Argentina's economy was particularly low, as the country was not only plagued by a growing current-account deficit, an overvalued currency, and an exchange rate pegged to the dollar, but also faced presidential elections in May 1995. In January 1995, Economy Minister Domingo Cavallo disbanded several government agencies in order to cut spending and met with the IMF, but he publicly argued that Argentina's 1991 law requiring the government to maintain foreign currency reserves to back it's currency precluded a Mexican-style monetary crisis.

Argentina's scheduled May 1995 presidential elections fuelled the economic uncertainty in the country, despite the fact that President Carlos Saul Menem, who altered the Constitution to permit himself to stand for re-election, was the leading candidate to win. The emergence of a new left-wing party, the *Frente Grande,* reflected the disappointment of Argentina's voters with both the Radical Civic Union and Menem's *Justicalista* party, as well as anger over corruption, the country's 12% unemployment rate (the highest since records began in 1970), and the economic deterioration of many provincial governments. Privatisation met little resistance in 1994: Argentina finished selling off most of its utilities, and the sale of the Comisión Nacional de Energia Atomica, which mines uranium, produces nuclear fuels, operates nuclear power plants and conducts research into nuclear applications, began. Privatisation of the post office, airports and petrochemical plants was under negotiation. Reform of the military was also a policy target. Since 1990 President Menem has cut military spending in half, reduced the size of the armed forces, sold off military-owned enterprises and, most recently, abolished compulsory military service in favour of a professional force. At the same time, he has also granted broad pardons to senior military officers on trial for human-rights abuses and to those who participated in past uprisings.

Although the Argentine economy grew at 6% in 1994, and inflation continued to fall to 4%, the growing current-account deficit, the reduction in capital inflows, and the country's increased interest rates suggest that recession will be on the policy agenda after the May elections. Increasing imports boosted the country's trade deficit to a disconcerting $6bn in 1994, although export earnings surged by 18%. The increase was due largely to trade with Brazil, particularly in the automobile sector, in anticipation of the customs union, Mercosur, between Argentina, Brazil, Uruguay and Paraguay that was signed in January 1995.

Brazil's economic policy trajectory faced similar scrutiny after the Mexican debacle. Brazil's new President Fernando Henrique Cardoso, a former left-wing academic turned finance minister, and the clear winner in a face-off against left-wing candidate Luiz Inacio Lula de Silva in November 1994, was the author of the country's July 1994 economic stabilisation programme launched under the previous Itamer Franco administration. The introduction of the new currency, the *real*, which was linked to the country's foreign reserves, slowed inflation to 25% (after it peaked at 2,500% in early 1994). This boosted confidence in the economy, led to increased foreign investment and a GDP growth of 4% for 1994. The newly overvalued currency, however, produced a trade deficit for the first time in five years. Even though domestic production and exports remained strong, Brazil acquired a new vulnerability to Argentina's emerging economic difficulties because of its increased trade with that country. In 1994, privatisation initiatives died under the interim Franco government, and congressional debate over oil and telecommunication sales stalled amidst union and student protest. Nevertheless the sale of electric power companies was approved, and shares of the state aerospace company were successfully sold.

Despite losing the presidential election, Lula de Silva's Workers' Party (PT) won two governorships in the September elections and doubled its representation in congress. Together with the powerful Brazilian Democratic Movement Party (PMDB) and other factions, they effectively blocked initial attempts by the Cardoso administration to push through legislation. The deadline for reforming Brazil's constitution passed in June 1994 with no agreement. Drug activity in Brazil increased in 1994, particularly in the poor urban areas, leading to a military operation in a slum near Sao Paulo after drug traffickers raided a nearby Brazilian naval base.

Nations of the West Coast

The Chilean economy did not feel the aftershocks of the Mexican financial earthquake: its ten-year history of solid export performance and careful regulation of financial flows contributed to a steady GDP growth of 4%. President Eduardo Frei, inaugurated in March 1994, did not alter the country's economic course, although the inflow of foreign capital forced the government to revalue the peso in November 1994, and inflation only just fell below 9%. The privatisation of state-owned assets was nearly completed as minority shares were sold in power utilities. Continued progress on constitutional reform was made in dismantling the constraints left behind by the Pinochet dictatorship, including recasting municipal government and legislative practices. The country's economic progress continued to be a model for the region, and in December 1994, NAFTA members agreed to negotiate for Chile's entrance into the regional trading bloc. Although such expansion of NAFTA was welcomed in Latin

America, most of Chile's exports already entered the region at zero or low tariffs, and its small economy would have a marginal effect on the regional agreement.

In 1994 Peru's GDP grew by a staggering 12%, the highest in Latin America, and inflation fell to under 15%, from a high of 7,650% in 1990. These excellent economic results further solidified President Alberto Fujimori's lead in the presidential elections scheduled for April 1995. Although Fujimori was accused of manipulating the country's election rules, reminiscent of his authoritarian acts of disbanding Congress in 1992 and rewriting the Constitution to allow himself to stand for re-election, his challengers did not pose a serious threat to his re-election campaign. In response to Peru's permissive foreign investment code, the country's four largest copper mines were successfully privatised in 1994, and the volume of foreign capital entering Peru more than doubled. Peru sold numerous small companies and allocated shares to the public in privatised utility companies, although banking, steel and oil companies remained under state control.

Unions protested at plans to break up and privatise the state oil company and applied for an injunction against the sale. High unemployment, increased income inequality, and regional and racial tensions soured the success of the region's most liberal free-market model. While the violent activities of the *Sendero Luminoso* guerrillas subsided in 1994, the increased drug operations in the country, and evidence of links between drug traffickers and Peru's senior military officers, raised new concerns.

The January 1995 territorial dispute between Peru and Ecuador, while boosting the political fortunes of both countries' presidents, was costly financially and in terms of human life; Peru admitted losses of several aircraft and helicopters, and Ecuador spent over US$10m a day during the 18 days of fighting. By early March, the attacking Peruvian Army, twice the size of Ecuador's, racked up 46 dead while Ecuador reported 30 casualties with at least 300 wounded. The United States, Argentina, Chile and Brazil, the guarantors of a 1942 peace treaty that had established the borders, negotiated several cease-fires, but no resolution to the conflict was reached. At the end of March more than 10,000 troops remained massed in the contested Amazon area, and the possibilities of a further flare-up were still high.

Ecuador's political system faced legislative gridlock in 1994, as President Sixto Duran Ballén faced a divided Congress and confronted a series of ministerial impeachments. The border conflict with Peru boosted his popularity, despite the heavy economic costs it inflicted on Ecuador. This new-found popularity, however, did not protect Duran Ballén from the strong resistance mounted by the Ecuador Congress, the unions and indigenous groups to his plans for privatisation, constitutional reform and a new land-reform law to break up communally held lands. Equally serious from his point of view, the conflict with Peru has boosted the popularity of Defence Minister General José Gallardo Román, who has not tried to hide his ambitions and who may decide to run for president next year. He would make a formidable candidate.

Colombia's Liberal Party candidate Ernesto Samper was elected President in the August 1994 contest. A more disturbing development was the election of several guerrilla-backed candidates in the remoter areas. Perhaps the only good news during the year was that Colombia's economy grew by almost 5% in 1994, and although inflation reached 22%, the government signed its first agreement with employers and unions to keep wages down. Yet even as President Samper opened an effort to achieve peace with the guerrilla organisations, and promised to make money-laundering an illegal offence, drug cultivation in the country increased, moving from coca to opium poppies for heroin. The US Congress considered denying Colombia aid and other economic privileges for failing to cooperate in fighting the drug trade. A decline in trade with a troubled Venezuela and a fall in coffee production led to reduced exports, but new oil and gas production figures presented a more optimistic future.

Venezuela was one of the few countries of Latin America (the Caribbean countries of Haiti and Cuba were the others) whose economy contracted in 1994, with GDP down by 4% and inflation standing at over 70%. Venezuela's President Rafael Caldera Rodriguez had taken office in February 1994 and was immediately confronted by the collapse of the Venezuelan *Banco Latino*, followed quickly by that of eight more financial institutions. Despite a massive injection of funds and extensive restructuring of the system the banks' difficulties led to the closure of the foreign exchange market in June 1994.

President Caldera was extremely active during the period. He imposed exchange and price controls and used new economic emergency powers to jail bankers, businessmen and financial analysts for their role in the financial collapse. In February 1995, as three more banks closed down, he threatened those that were still solvent with nationalisation unless they recapitalised. Although President Caldera began his term vowing to slow down market reforms, his administration opened more state oil and gas monopolies to foreign participation, and offered more incentives to encourage foreign capital.

Small is not Necessarily Better

All the smaller Latin American economies managed steady growth of 2–4% and maintained inflation rates at 10–45%, showing only sporadic increases in export earnings, largely from natural resource sales. Uruguay's Colorado Party candidate and former President Julio María Sanguinetti narrowly won the November 1994 presidential elections, and the sizeable Congressional representation of the *Blanco* and the left-wing *Encuentro Progresista* opposition parties presented legislative difficulties in early 1995. In Paraguay, the opposition-led Congress took greater control over the choice and manner of state company privatisations; the sale of telecommunications, water and electricity remained on the agenda despite widespread strikes and protests. Paraguayan peasants, who make up 50% of the country's population, also clashed with police over land reform. Tension between the military and Congress ensued over Paraguayan President Juan Carlos Wasmosy's call to dismiss the commander of the army, both over the proposed law barring the military from participating in politics, and over the military's role in the drug trade.

Bolivia's President Gonzalo Sanchez de Losada also battled an irascible Congress, which was encouraged by street protests to hold up his free-market economic policies. His administration's tough anti-drug campaign also encountered widespread opposition from unions. Bolivia's resentment of the US involvement in the anti-drug effort increased sharply when the US alleged that former President Jaime Paz Zamora's cabinet had been involved in drug trafficking.

The Central American economies all grew at respectable rates of 3–4%, and inflation was well under control, with the exception of Honduras, which recorded no growth. Presidential elections scheduled in Guatemala for November 1995 will likely pit the unelected President Ramiro de Leon Carpio against former military ruler Efrain Rios Montt of the Guatemalan Republican Front – if Montt succeeds in circumventing a constitutional ban on his candidacy. A peace agreement between the Guatemalan government and the United Revolutionary Guatemalan (URNG) guerrilla movement was not signed as scheduled by the end of 1994. In February 1995, El Salvador's President Armando Calderón Sol announced a free-market economic programme that reduced import tariffs on capital goods to 1%, and adopted a currency programme similar to the Argentine system, which pegs the currency to the US dollar and permits free convertibility.

Cuba faced another exodus of its people in summer 1994; thousands of Cubans sailed by rafts to US waters and were transferred to makeshift camps in the United States, Panama and in US-occupied Guantanamo Bay in Cuba. After President Clinton announced in August that Cuban refugees no longer qualified for automatic entry into the United States, the two countries agreed that, in exchange for halting further illegal migration, the US would increase its annual quota of legal immigration to 20,000 persons. While this issue was put to rest through negotiation, there were no talks on the broader issue of the US embargo on Cuba, although there were tentative indications in early 1995 that some in the White House were giving consideration to an adjustment in the long-standing policy.

There were signs in 1994 that Cuba is beginning to recognise that its economic model will have to be changed if economic catastrophe is to be kept at bay. Faced with declining industrial output, poor sugar harvests and high budget deficits, Castro began to take some steps towards reform, although publicly he and Cuban officials continued to extol the planned, socialist nature of the economy. Self-employment was legalised, public markets were authorised, joint ventures with foreign firms were sanctioned, cuts in subsidies were made to loss-making state companies, and new taxes were imposed. A senior IMF official visited Cuba twice in 1994, a signal that Cuba was looking towards the world economic community for help. In December 1994, the Cuban government formally introduced a convertible currency to serve as a foreign-currency certificate to be exchanged at a 1:1 rate with the US dollar, which lowered the open black market rate to 40–50 pesos to the dollar, down from 120.

Haiti, newly returned to democracy by the peaceful invasion by US forces in October 1994, postponed its legislative elections until June 1995. President Jean-Bertrand Aristide's new administration was unable to lift Haiti's economy out of the devastating effects of the international economic blockade. In December 1994, Haiti received $40m from the World Bank, its first new international lending since the military overthrew President Aristide in 1991.

Regional Economic Integration

Trade liberalisation in the form of regional integration remained high on the political agenda during 1994–95, as NAFTA, Mercosur, the Andean Group, and the Central American Common Market, as well as numerous bilateral trading partners, signed new treaties. The hemispheric summit held in Miami in December 1994 was heralded as a great success. Leaders from Latin America, the Caribbean, the US and Canada agreed to establish an American Free Trade Area (AFTA) by 2005. Nevertheless, trade between these various subregional groupings in Latin America was still on a downward slope, and over 60% of export earnings in the region still comes from commodities such as foodstuffs, fuels and minerals.

On 1 January 1994, NAFTA, which groups the US, Canada and Mexico in a free-trade zone, came into effect. In the first six months, Mexico and Canada increased their exports to each other by 36%, and trade between Mexico and the United States increased by 20%. Over the course of the year there was a shift in trade flows; for the first six months, Mexican exports to the United States grew faster, while in July this trend reversed.

The Mexican textile, clothing and shoe industries were most threatened by US exports, while Mexico's agriculture sector, aided by continued Mexican government subsidies, escaped the full effects of NAFTA's open trade policies. Mexico's newly created Federal Competition Commission was successful in fining companies that fixed prices or manipulated markets, although the resolution of trade disputes among NAFTA members was contentious, especially in the areas of tuna, steel and cement. In this first year, the effect of NAFTA on job creation and international labour rights remained unclear.

The trade agreement led to some coordination of financial issues. In April, the US, Canada and Mexico set up a multi-billion-dollar swap arrangement to help cushion sharp fluctuations in foreign exchange markets. While Canada was able to use these funds to good purpose, the December 1994 peso devaluation was far too large for this fund to handle. The devaluation of the peso promised to open Mexico to even more foreign direct investment, particularly in the oil industry, and to boost Mexico's exports. Partly in response to the US executive-led Mexican bail-out and to fears of increased export competition, some in the Republican-led US Congress began mobilising to repeal NAFTA in early 1995.

In March 1994, representatives of 22 countries and dependent territories in the Caribbean basin concluded their first round of talks on gaining membership in the Association of Caribbean states, an organisation aimed at liberalising investment, coordinating negotiating positions in bilateral and multilateral issues, and promoting investment opportunities. In June leaders of Latin America's 19 largest nations met to discuss the future of regional integration and free-trade areas, in particular how to merge the different integration schemes. In July, 13 member-countries of the Caribbean Community (CARICOM) discussed applying for membership in NAFTA, since displacement of Caribbean exports by NAFTA is expected to run at $35–53m annually.

Tariffs on most goods traded among the four member-countries of Mercosur have been gradually reduced since 1991, and in January 1995 Argentina, Brazil, Paraguay and Uruguay formalised this trend by entering into a customs union. The countries agreed to end duty-free imports of capital goods and imposed a 0–20% tariff on all imports, although they continued to give special treatment to cars, wheat, sugar and textiles as well as protection for high-tech goods. In December 1994, the EU and Mercosur agreed to begin negotiations on a common free-trade zone. The EU has been the region's largest trading partner since 1986, and Mercosur was the fastest growing market for EU exports. In January 1995, Costa Rica and Mexico agreed to eliminate tariff and non-tariff barriers on 12,000 products over a ten-year period.

Attention Must be Paid

Latin America covers a huge area and its countries, their governments and their economies are characterised by diversity. Generalisations are thus difficult to make. Yet many of them have in common a recent past of authoritarian military rule which they have cast off. Now fledgling democracies, they are struggling with poverty, deprivation and, in many cases, the strong remnants of feudal social fabrics in their efforts to create fairer political and economic societies. The collapse of the Mexican peso has deepened the task they face. Unless the stronger developed industrial democracies are willing to encourage their admirable efforts with deeds instead of just words there is a serious chance that they will not be able to reach the desired goal of democratic societies whose economies are run according to free-market principles.

Europe

The progress of political and economic reforms continued to be the major preoccupation throughout 1994 and early 1995 in the former Soviet bloc, while Western Europe was enlarging and adjusting its institutions. In the early part of 1994, Russia seemed on a track to greater normality, but in October the rouble collapsed creating economic turmoil. This disaster was compounded by the bungled Russian intervention in the breakaway republic of Chechnya. While the Chechen affair may have served to discredit the party of war inside Russia, it has not helped Russia's image abroad – and as a result, may make the West even less likely to listen to its objections over North Atlantic Treaty Organisation (NATO) expansion. In Ukraine, however, things looked better at the start of 1995 than they did at the start of 1994. After much prevarication, Ukraine finally decided that to better its relations with the West – and therefore receive much-needed assistance for its internal economic reforms – it had to respond to Western requests to join the ranks of the non-nuclear states.

In Western Europe, the European Union (EU) busied itself with the adjustments needed to enlarge by including Austria, Finland and Sweden in its ranks. It also prepared to accept the East European states at a much later date. Enlargement raises inevitable issues of 'deepening', or further integration, which will have to be dealt with at the Inter-Governmental Conference (IGC) scheduled for 1996. Also on the table will be proposals to reform the current variable geometry of overlapping security and defence organisations. With NATO beset by transatlantic tensions over the never-ending Bosnian crisis, most member-states seem willing to strengthen the Western European Union (WEU).

The Bosnian crisis remains a large blot on the West's record of managing crises in the post-Cold War world. The terrible pattern of rising hopes for peace being cruelly dashed by continuing fighting on the ground repeated itself in 1994 and 1995. The major multilateral organisations, the United Nations (UN) and the EU, were shunted aside as a concert of five great powers stepped in to try to resolve the crisis, but with no more success than its predecessors. There was one very bright spot in Europe, however: the guns have been silenced in Northern Ireland for months and significant high-level peace talks could soon begin. If this long-standing conflict is resolved peacefully, it would rekindle hope that it is possible to find a political solution for other disputes which now appear as intractable as that in Northern Ireland always has.

The Struggle for Normality in Russia

In late summer 1994 Russia seemed to some to be on the point of a breakthrough to stability and normality, the two things that opinion polls consistently show Russians want more than anything else. The constitution, which had been adopted by a referendum held in December 1993 at the same time as the parliamentary elections, has set new rules for the game of Russian politics. The crude division of powers between the federal and local governments set out in the constitution helped to defuse long-simmering rows about tax revenues and thus to reduce secessionist trends, which at certain points in 1992 and 1993 had looked ready to pull Russia apart. The new parliament proved not to be nearly as bad as people had initially feared. Even the Communist and Agrarian Parties spent much of the first nine months of 1994 looking for ways in which to build a constructive relationship with the President and the government.

The growing maturity of Russian politics rested on two foundations. The first was scattered indications that the pace of economic collapse was slowing. By the third quarter of the year, there were tentative signs that the slump in production was reaching an end. Monthly inflation ran to single digits for six months and levelled out at 4.4% in August. The rouble appreciated against the dollar in real terms, helping to increase the purchasing power of the average industrial wage measured in dollars from $8 per month in January 1992 to over $110 a month by mid-1994.

The first stage of mass privatisation, which finished in June 1994, had placed over 14,000 medium and large enterprises in private hands, turning 40 million Russians into shareholders and boosting a private sector which created 62% of Russia's gross domestic product (GDP) in 1994. By August 1994 foreign portfolio investment was flowing into Russia's emerging stock-market at the rate of $500m a month.

The second foundation was Russia's growing confidence in its ability to defend its interests beyond its borders. One by one the other members of the Commonwealth of Independent States (CIS) were being brought back to the table with Moscow, and in some cases back into its sphere of influence. Georgia was forced into signing an agreement accepting Russian military bases on its territory with no date set for their withdrawal. In Azerbaijan, Geidar Aliyev received a nasty shock in September 1994 when opposition forces, with backing from Russia, nearly toppled him. This was Russia's way of reminding Aliyev that Azerbaijan's oil resources can only be exploited with Russia's consent and participation. The 14th Army remains in Moldova and shows no sign of moving, although a withdrawal agreement has been signed. Alexandr Lukashenko was elected President of Belarus in large part because he promised to reintegrate his country with Russia. The Russians, however, have refused to reintegrate Belarus unless and until it puts its economic house in order – at which time it may not need integration. Further afield it became clear that while Russia may not have a veto on how NATO might be expanded, its involvement in and influence on that process will be considerable. Thus, by autumn 1994, Russians may not have had cause to feel content, but they did at least have good reasons to be optimistic about the future. Then everything went wrong.

First, on 11 October, the rouble collapsed in value. Then President Boris Yeltsin decided to re-establish Moscow's writ in Chechnya by force. The disastrously botched war in Chechnya has done immense damage to Yeltsin's image at home and abroad. Had there been a presidential election before the crisis in Chechnya, Moscow intellectuals would have grumbled about the dangers of re-electing a man who had drunkenly conducted a band in Berlin and been too tired to walk down the steps of his plane to exchange greetings with the Irish Prime Minister, but Yeltsin would still have been the odds-on favourite to win it. The Chechnyan war has made Yeltsin's re-election less likely, but there is still no obvious alternative.

Thus Russian politics has entered a new era of uncertainty. The economy is once again in a mess: inflation was running at 17.8% a month in January 1995, and foreign investment has dried up. At best, Russia has thrown away a year's progress in its struggle to become a 'normal' country. At worst, the way has been opened for an extreme populist to triumph in the presidential elections to be held in June 1996.

Aftershock from an Election

The foundations on which Russia's progress in the first part of 1994 were based were undermined by the delayed impact of the December 1993 elections. Russia's Choice, the largest pro-reform party, had emerged from these elections with the greatest single block of seats in the state Duma (the lower house in Russia's new parliament), but it still performed below expectations. It won only 16% of the votes cast for the 225 deputies elected by proportional representation from lists put forward by political parties. Vladimir Zhirinovsky's neo-fascist Liberal Democratic Party won 22% of the votes cast for these seats, but fared worse than Russia's Choice in the 225 single-member constituencies. The Communists and Agrarians also polled strongly.

To some extent Yeltsin was to blame for the reformers' poor showing. The founders of Russia's Choice had expected their party to receive the President's full support. Instead Yeltsin stood aloof from the election campaign, reasoning that it was safer for him to retain as much room for manoeuvre as possible so that he could negotiate with whoever won.

The disappointing election results meant that Yeltsin had to use his room for manoeuvre to make a strategic decision. He could either see the election as proof that he had to use the powers granted to him by the new constitution – powers greater than those wielded by almost any other elected president of a major country – to push more energetically towards a free market, democracy and the rule of law. Or he could go along with those who believed that Russia needed a larger state, more bureaucratic controls and more aggressive foreign policies. That is, he could adopt the policies espoused by Zhirinovsky in order to defeat him.

Yeltsin did not make this decision in a hurry. Instead, in the first part of 1994, he largely disappeared from public view, leaving the running of the government to Viktor Chernomyrdin, who had replaced Yegor Gaidar as prime minister in December 1992. Chernomyrdin had made his career at *Gazprom*, the Russian gas monopoly which supplies 40% of Germany's natural gas and could one day become one of the world's largest private energy companies (it was privatised in June 1994). Thus Chernomyrdin came into office with a reputation as a tough and effective manager. In some areas, Chernomyrdin has played a useful role. He has made considerable efforts to draw the major military industrial ministries into the privatisation and economic reform circle, and has worked hard to lower barriers to foreign trade and investment. In general, however, his record of achievement as prime minister has not been good, for three reasons.

First, the government has no political base. Attempts to push a vote of no-confidence through the state Duma have, so far at least, failed to attract a majority (226 votes). Yet the government can only safely rely on the support of 80–90 deputies. Chernomyrdin's second problem throughout 1994 has been his increasingly tense relationship with the President, although some of the strain lifted in early 1995. Yeltsin seems to feel threatened by having a powerful and effective prime minister and has tended to undermine him on the few occasions that Chernomyrdin has taken the initiative to develop and push through policies. The third handicap is one of the prime minister's own making – he in turn appears to feel threatened by having a finance minister with a grasp of macro-economic policy (which the prime minister lacks). He used the disappointing performance of the reformers in the elections as an excuse to ease Yegor Gaidar and Boris Fedorov, two of the most prominent reformers, out of government. Despite their departure, inflation continued to fall in early 1994. This was mainly because Fedorov, Finance Minister until January 1994, had kept a tight grip on the new money supply in the last quarter of 1993.

In April 1994, Chernomyrdin's government began to increase the money supply which rose by roughly 10% a month in the second and third quarters of the year. Increasingly the Central Bank was forced to intervene to support the rouble against the dollar. By the end of September inflation was rising again, the Central Bank was running out of reserves and the government seemed blithely unaware of the impending consequence. On

11 October, the inevitable happened: the rouble crashed, losing two-fifths of its value in one day.

The Ascendancy of the Party of War

Politically, the collapse of the rouble could not have happened at a worst time. Exactly a week earlier, President Yeltsin had used a nationally televised press conference to mark the first anniversary of the dissolution of the old parliament as a platform to herald the arrival of a new era of financial stability in Russia. After the rouble's collapse, Yeltsin wanted to find who was to blame for making him look stupid.

The inquest into who had tried to topple the President by forcing down the rouble was conducted by the Security Council. This shadowy body, under the leadership of Oleg Lobov, an ally of the President from Yeltsin's days as party chief of Sverdlovsk, has taken on some of the functions of the old Politburo.

As the Security Council's investigation got under way a scandal put the representatives of the power ministries, who dominate the Security Council, onto the defensive. On 17 October Dmitry Kholodov, a young investigative journalist, was killed by a bomb planted in a briefcase which a contact in one of the power ministries had arranged for him to collect. Moscow newspapers all concluded that he had been killed to prevent the publication of a series of articles which would have charged General Matvei Burlakov, commander of the army's Western Group of Forces (based in Germany until August 1994), with massive corruption.

The allegations, if substantiated, would also have ended the career of Defence Minister Grachev, a close colleague of General Burlakov. To retain his job and avoid a criminal investigation, General Grachev had to get something right. A short and successful colonial war seemed the answer.

This is one of the few plausible reasons why the Security Council decided to use force to re-establish the rule of law in Chechnya, which had announced its secession from Russia three years earlier. Attempts by the Russian government to topple the regime of Dzhokar Dudayev, first by supporting the Chechen opposition and then through covert operations, had made little progress by September. But this did not explain why two months later Yeltsin decided to invade. Russian covert operations in other parts of the Caucasus, notably in Abkhazia and Azerbaijan, have been successful. In addition, in Chechnya time was on Moscow's side. Because Dudayev comes from a small Chechen clan, the numbers of armed men he could automatically count on was small. Under his rule the social and economic structure of Chechnya had collapsed as unemployment rose to over 50% and it became by far the poorest part of the Russian Federation. Although Dudayev was on the verge of falling like a rotten fruit, the Russian government did not have the patience to wait.

Yeltsin chose not to wait for at least two reasons. Chechnya is an important centre for Russia's oil refining and distribution system. The refinery at Grozny is the second largest in Russia, with the capacity to

process 18 tons of oil products a year. More important, Chechnya sits astride the pipeline which Russia hopes will carry most of the oil from new fields around and under the Caspian Sea. Russia is determined not to let Azerbaijan slip from its grasp, which it might do if it were able to export its oil south through Iran, or west via Turkey. In order to convince the foreign companies and banks which will finance the development of these new pipelines that the safest route lies through Russia, Chechnya had to be brought back under control.

Second, Yeltsin believed Defence Minister Grachev's blandishments that the whole operation would be over in days. He wanted the Russian armed forces to carry out the equivalent of the American re-imposition of order in Haiti to demonstrate that Russia is still a power with which to be reckoned, and thus to strengthen his support amongst nationalists.

The Debacle

The invasion of Chechnya had little in common with the American invasion of Haiti, and was in the event extremely humiliating. The first direct military intervention on 12 December 1994 was half-hearted and almost immediately ground to a halt. This gave the Chechens time to prepare for Russia's main assault on Grozny, launched on New Year's Eve.

The Russians had bothered to work out only the crudest operational plan. Three armoured columns were supposed to enter Grozny from the west, north and east, while an escape route was left open to the south. This was intended to allow civilians to leave and enable Russian air power to wipe out Chechen military units as they later fled the city. The exact opposite happened. The commanders of the eastern and western columns refused to move on the city, leaving the first stage of the fighting to the northern column. Its armour almost immediately became separated from the supporting infantry. The surviving tank crews later complained that they had been sent into Grozny with unclear orders, no maps of the city and radios that did not work.

Some of the conscripts had hardly been in an armoured vehicle before. They were sitting ducks for the highly motivated Chechens. United in self-defence and well-equipped with anti-tank weapons they were regularly reinforced through the southern corridor. The Russian thrust did not turn into a rout, however, largely because of the leadership of General Lev Rokhlin, the only soldier to emerge from the Chechen operation covered in glory. In the meantime, General Dudayev has been transformed from the increasingly hated leader of a criminal gang which was looting Chechnya into the heroic leader of its national resistance.

Despite the incompetence of the first attack, it seemed inevitable that Russian might would triumph in the end. What was surprising was how long it took for the Russians to control Grozny – almost two months, and at the cost of razing much of the city. In late March 1995 heavy fighting was still taking place along the Argun River to the south-east of the city.

Beyond the Debacle

Russians now joke that modern Russian politics can be divided into two eras: BC (before Chechnya) and AD (after disaster). Nevertheless, of the four major consequences of the war in Chechnya, three could yet be helpful to Russia's struggling attempts to build a normal country.

The first consequence is the discrediting of the party of war. This group promised a quick military victory, which it singularly failed to deliver, and then compounded its failure by insisting that negotiations with Chechnya could be on only one subject: the Chechens' unconditional surrender. Anyone with the slightest knowledge of Caucasian history will know that this is an oxymoron. Thus, having dragged Yeltsin into an unwinnable war, the party of war then offered him no way out.

Considering the damage this has done to the President's popularity, his vengeance has so far been limited. Nikolai Yegorov, the deputy prime minister responsible for nationalities and a hardliner from Krasnodar, took to his hospital bed, the traditional refuge for a Russian leader in disgrace. Oleg Soskovets, another deputy prime minister and the war party's favourite to replace Chernomyrdin as prime minister, has been relieved of most of his economic duties and been given responsibility instead for sorting out Chechnya, a poisoned chalice which many observers assume is the prelude to his political demise.

Yet General Grachev has managed to remain as Defence Minister, in large part because Yeltsin fears that any replacement acceptable to the army would be less loyal to the president. General Grachev's loyalty is guaranteed by his unpopularity within the armed forces. And Alexander Korzhakov, the President's chief bodyguard, closest political ally and leading figure in the party of war, has managed to maintain his position despite the difficulties over Chechnya. Nevertheless, the standing of the war party has been much weakened by the conflict.

The second consequence is the increasing self-assertion of Russia's regions. None has yet tried to follow Chechnya and attempted to secede from the Federation. Many local governments are, however, grabbing more powers from the federal government, on the basis that its behaviour in Chechnya proves it is not yet to be trusted. As long as this further devolution of power to the regions takes place in an orderly way (within the framework laid out by the constitution and the new law on self-government that the President will present to parliament this spring), the result is likely to be better government in much of Russia.

The third consequence is that the war in Chechnya has brutally exposed the state of the Russian Army. The ground troops lost as many as 200 armoured vehicles in the first major assault on Grozny because their crews had no radio contact with supporting infantry, who mostly refused to advance. In many cases the drivers had little training and no maps of the city. This lack of preparedness reflects the chronic undermanning of the Army. For example, 84% of the conscript pool is more or less legally entitled to deferment or exemption. Of the remaining 16%, half are sent to the Interior Ministry and FSK (one of the successors to the KGB), leaving only 75,000 conscripts a year for all the armed forces. One-third of these dodge the draft, safe in the knowledge that the courts are unlikely to pursue them. The quality of those who do end up in the Army is abysmal. Often illiterate and addicted to drink or drugs, one-fifth of the conscripts have criminal records. Such an underclass may have enabled Wellington to win at Waterloo, but is not much good at driving tanks.

The Army is likely to become impotent before it improves. The US Army managed to overcome its post-Vietnam trauma by becoming an all-volunteer force with a decent non-commissioned officer (NCO) corps on which to build and a dedicated officer corps. The Russian Army cannot afford the first solution, shows no sign of being able to overcome its long-standing lack of experienced NCOs, and has a disillusioned officer cadre. Junior and middle-ranking officers know that their superiors are most interested in business deals. There is virtually no money for training or maintenance of equipment; Alexander Lebed, the Army's leading dissident general, estimates that only one-fifth of the tank fleet is usable. Pilots receive little systematic training and log only 25–30 hours a year (the NATO minimum to maintain efficiency is 120–130 hours).

One result of the Army's abysmal performance in Chechnya is encouraging: the fear that Russia's armed forces could help to rebuild the Soviet Union through force now looks risible. The other consequence of the armed forces' humiliation in Chechnya is more disquieting. The use of the armed forces to solve an internal problem has inevitably dragged the Army deeper into politics. The first major step in that direction was in October 1993, when Yeltsin eventually managed to persuade a handful of troops to shell the old parliament.

At the time this looked like a regrettable but necessary act to remove a legislature which was threatening to take away most of the President's powers and reverse what reforms had taken place. But, with hindsight, it is now clear that the use of tanks in Moscow placed Yeltsin deeply in debt to the armed forces, making it impossible for him to push ahead with the army reforms he has repeatedly promised.

This failure has produced the current dispirited force, many of whose officers now argue that if they are forced to do dirty work inside Russia, they might as well do it themselves and not at the behest of incompetent civilian politicians. Even if Chechnya and the shelling of parliament have not made a military coup more likely, they have reintroduced armed force as a major factor in Russian politics.

Who Comes Next?

It still seems probable, however, that the single most important question in Russian politics will be decided democratically: who will win the presidential election scheduled for June 1996? The answer, to a large extent, will depend on what happens to the economy. Recently commissioned opinion polls show that while 30% of Russians think that Chechnya is the most important problem for the government to solve, 90% named inflation as the most important problem which must be solved if life is to become normal.

A sharp reduction in the rate of inflation in 1995 is technically quite possible. Inflation in 1992 and 1993 was fuelled by three sources: lending from Russia to other members of the CIS; subsidies to industrial enterprises; and printing money to finance the government's budget deficit. By the end of 1994 two of these sources of inflation had been blocked – lending to the CIS and across the board to industry had stopped. The draft budget for 1995, which passed its fourth and final reading in the lower house of parliament, sets a cap of 73 trillion roubles on this year's budget deficit. That is equal to 5.6% of forecast GDP. The government will be able to finance two-fifths of its deficit with a $6.4 billion loan from the International Monetary Fund (IMF), agreed in mid-March. The rest of the deficit will be financed by selling treasury bonds. The older method of deficit financing – just printing money – was made illegal by a law passed at the end of January 1995.

The first convincing sign that the government can live with these constraints is that the money supply did not expand in January or February. This means that by the end of the year the government should be able to reduce inflation to around 2% a month, down from 17.8% in January 1995. The decline in production appeared to have levelled out in the fourth quarter of 1994. That, combined with lower inflation, should cause saving and investment to increase as the year progresses.

If the government does not lose its nerve, the presidential election could be won by an insider, who could credibly claim the credit for this achievement. That would mean a field of leading contenders including Chernomyrdin, Ivan Rybkin (the speaker of the Duma), Anatoly Chubais (the privatisation czar who now has overall responsibility for macro-economic policy) and, of course, Yeltsin, as long as he can prove that his drinking is under control and that he is physically up to a second term in office.

What if the government again botches stabilisation and the election takes place when inflation is still high and economic growth a distant prospect? The chances then of an outsider winning the presidency are much higher. Russia's size means that the campaign will to a large extent be fought on television. Under these conditions a populist with a gift for sound-bites and the money to pay for television time, but without necessarily a large party apparatus, could run an effective campaign with relatively little preparation. At the 1993 parliamentary elections, one unquestioned outsider campaigned – Vladimir Zhirinovsky. In 1996 he is likely to face stiff opposition from other demagogues. They include: Alexander Rutskoi, the former vice-president; Alexander Barkashov, Russia's best-organised neo-fascist; Pyotr Romanov, an unreconstructed communist; and one-time reformer Boris Fedorov, the former finance minister whose turn of phrase, while being economically literate, can sometimes take on a nationalist accent.

The other place to look for a presidential candidate from outside the current political mainstream is in the armed forces. A general could do well with older voters who feel threatened by Russia's loss of status in the world, while at the same time picking up votes from 'new Russians' who have benefited enough from reform to want a Pinochet-like figure to protect their gains by enforcing law and order and reducing taxes. General Lebed is the most likely soldier to make a serious bid for the presidency.

If he, or any other of the populists, were to win the presidency, their campaign rhetoric would probably force them to re-introduce price controls and to pump out more subsidies in order to increase production. The market's reaction would cause a complete collapse of the rouble. And, in practice, price controls would only be enforceable in today's Russia if the government were willing to shoot the first businessman caught raising prices.

Many liberals who supported Yeltsin until the crisis in Chechnya argue that the consequences of that war have made the return of an authoritarian state – which would not think twice about shooting its own citizens – more likely. Their fears were heightened when a bill to return many of the powers once enjoyed by the KGB to the present security services sailed through parliament, as well as by Yeltsin's refusal to purge the leaders of the party of war.

Reality is probably slightly different. The power ministries are so debilitated that they could probably only make a dictatorship work in Moscow and St Petersburg, while the rest of the country went its own way. Even in those cities it might be hard to cow the population. While many of the other features of a healthy civil society have yet to appear in Russia, the country does now have a vigorous and largely free press. As important, the new entrepreneurial class is dominated by young people. The Soviet system was already in decline when they left school. They are used to operating in a tough and anarchic business environment, and it is hard to make them afraid. Without fear, a new dictatorship in Russia is doomed to fail.

The real risk is a subtler one. While any attempt to return to the past is almost bound to fail, just as it did in France in 1982 and in much of Eastern Europe in 1994, the question is how much damage could the attempt do? Russia is nowhere near as stable as Poland, let alone France. A failed attempt to reassert the state's power over the market and to make the entrepreneurs slaves of the state would set Russia's search for normality back for years, possibly for decades.

Ukraine: Rising from the Ashes

To many observers in late 1993, Ukraine was a state on the verge of sinking. Its economy was near shipwreck. Even hyperinflation in autumn 1993 could not rouse President Leonid Kravchuk to take steps towards reform. Strikes in the east in summer 1993 raised serious questions about the country's unity and forced the government to schedule parliamentary and presidential elections in 1994. Ukraine's shilly-shallying on nuclear disarmament deepened its isolation from the West. And its growing weakness and disarray had increasingly destabilised its relations with Russia.

Despite these difficulties, in the words of its own national anthem, Ukraine 'has not yet perished'. In 1994, the government surprised many observers by taking the first steps towards putting its own house in order and gaining international support for economic reform. It did so first by assuaging doubts about its nuclear intentions, signing the Trilateral Agreement with Russia and the US in January and acceding to the Nuclear Non-Proliferation Treaty (NPT) as a non-nuclear state ten months later. It elected a new parliament and president who launched an economic reform programme in October. Beyond this, the country continued to manage relations with Russia without either outright defiance or capitulation. By the end of the year, Ukraine had done much to shed its image as the sick young man of Europe.

Nuclear Disarmament

For Western states, the nuclear question overshadowed all other aspects of their relations with Ukraine. Within Ukraine itself, however, nuclear matters were never as central. There the problems of consolidating a new state, dealing with economic decline and creating a stable relationship with Russia loomed larger. By the end of 1993, the Ukrainian leadership understood that continued delay on the nuclear issue only impeded its ability to address these other core issues. In particular, Ukraine needed external political and economic assistance. Such outside aid depended in turn on the resolution of the nuclear question.

There were three major accomplishments on nuclear disarmament in 1994. In January, Presidents Clinton, Kravchuk and Yeltsin signed a trilateral agreement, committing Ukraine to transfer to Russia for dismantlement all nuclear warheads on Ukrainian territory. In return, Ukraine was compensated for the highly enriched uranium returned to Russia with fuel rods for its nuclear power plants, US technical and financial assistance to meet its disarmament obligations, and security assurances, which were only formally extended to Ukraine after it acceded to the NPT.

In the months that followed, the Ukrainian government faithfully fulfilled its obligations under this agreement. By the end of the year, it had transferred over 360 warheads, well in excess of the 200 that the Trilateral Agreement required by then. In November 1994, on the eve of President Leonid Kuchma's visit to Washington, the Ukrainian parliament voted to accede to the NPT as a non-nuclear state. President Kuchma passed on Ukraine's formal accession to Presidents Clinton and Yeltsin and UK Prime Minister John Major in December in Budapest. The four leaders also signed a memorandum formally extending a set of security assurances to Ukraine.

The breakthrough on nuclear matters immediately injected new dynamism in US–Ukrainian relations and in Ukraine's relationship with the West as a whole. Kravchuk visited Washington in March 1994 and was granted additional bilateral aid. The US took the lead in seeking Western economic assistance for Ukraine at both the July Group of Seven (G-7) summit in Naples and the October G-7 Donors' Conference in Winnipeg. Though no formal linkage existed between disarmament and support for economic reform, there is little doubt that a continued stalemate on nuclear weapons would have dampened US and Western enthusiasm for aid to Ukraine.

Internal Political Developments

Ukraine has a large Russian population (11,000,000 according to the 1989 Soviet census), mainly concentrated in the eastern regions. These regions are also home to the heavily subsidised mining and industrial enterprises that prospered under Soviet rule and remain heavily dependent on governmental support. Many feared that the economic decline would encourage regional fragmentation in the wake of the 1994 elections that could very well tear Ukraine in two. The actual story, however, was quite different.

The first rounds of the parliamentary elections in March and April 1994 gave the left-wing parties based in the east, such as the Socialist and Communist Parties, the largest bloc of seats (nearly 150 of the 336 seats decided in the first round and run-off). This bloc helped to shape the new parliament, and its success allowed it to elect one of its own members as Speaker. But the notion that the triumph of the left or of the eastern regions was complete has been modified by subsequent political developments in Ukraine.

In the first and successive rounds of elections, eastern Ukraine also sent its share of reformers and moderates, including Kuchma himself, to the parliament. In addition, the left-wing parties have no real programme of their own except to delay or block reform. Neither Kiev nor Moscow has the wherewithal to provide subsidies that would preserve the old industries. Subsequent rounds of parliamentary elections have further diluted the initial pre-eminence of the left wingers, and the emergence of an energetic president with a reform programme has shifted the political momentum to the executive branch.

The presidential elections appeared to provide additional evidence of regional fissures within Ukraine, although Kuchma's performance in office gives the lie to claims that he would compromise independence by turning to Moscow for salvation. Kuchma has provided considerable continuity in the executive, particularly in his dealings with Russia, except where a decisive break with Prime Minister Vitalii Masol was needed most

– on economic reform. Nevertheless, the presidential campaign and the subsequent voting patterns (Kuchma received less than 10% of the vote in the west and Kravchuk less than 25% in the east and south) underscores that regional divisions are there to be exploited.

But it would be wrong to exaggerate the meaning or significance of these divisions. The campaign itself exacerbated regional differences. Kravchuk, having little to offer in economics, tried to cast Kuchma as a lukewarm supporter of Ukrainian independence. Yet his attempts to evoke anti-Russian fears as though they were equivalent to Ukrainian patriotism cost him as many votes in the east as he gained in the west. Ultimately, Kravchuk's failure to dominate the middle regions of the country was his downfall. He lost some of these districts and won others, but by smaller margins than he needed. Here people voted against his economic performance, not for reunion with Russia.

The past year has exposed serious fault-lines in Ukrainian politics. Yet the ethnic, political and economic diversity of Ukraine may turn out to be a source of strength, preventing the domination of Ukrainian politics by either re-unionists or radical nationalists. Ukraine's diverse population could not be easily governed by either extreme and its stability requires the integration of Russian and other minorities into the country's political and economic life. Ukraine has developed as a political and territorial state, not as an ethnic one. In the first rounds of the parliamentary elections and in the presidential contest, Ukrainians went to the polls in large numbers. By so doing, the electorate demonstrated that it is looking towards Kiev for solutions to its problems and is giving Ukraine's fledgling political process a chance to get to grips with public concerns.

This cycle of elections altered the players in both the executive and legislative branches, but the key internal political battle remains one of power and policy, not personnel. This round of elections postponed the struggle for pre-eminence between the two branches of government. President Kuchma has proposed a 'law on power', modelled on the Polish constitution, that would provide the constitutional basis for a strong president. This law is expected to be put to the vote in the first half of 1995.

Beneath this constitutional issue, however, lie enormous differences within Ukrainian politics over policy on private property, market reforms and the evolution of democracy in Ukraine. The struggle will also determine whether many of the old guard are able to retain their power and influence. Thus, in the coming months, a key test of political reform in Ukraine will be whether Kuchma has his way on the law on power.

Crimea

The divisions within Ukraine that produce the greatest concern are most palpable between Ukraine proper and Crimea. Crimea is the only region of Ukraine where Russians form a majority. In January 1994, to Kiev's dismay, the Crimean electorate chose Yuri Meshkov, the candidate of a Russian nationalist party, as its first president. Meshkov promised during the election campaign to hold a referendum on Crimean independence. A war of decrees ensued between Kiev and Simferopol that transformed the referendum into a poll on autonomy. On 27 March, over 80% of Crimean voters supported the call for increased autonomy and nearly as many supported the notion of dual Russian–Ukrainian citizenship for the residents of Crimea. Many of these same voters refused to participate in the national parliamentary elections held at the same time, thereby sending a strong signal to Kiev of their demand for local autonomy.

Throughout the spring, there remained a real danger that the Crimean crisis would escalate as Meshkov pressed claims for control over local police and military forces. However, four factors kept the crisis in check until early 1995. The Russian government has kept Meshkov and his followers at arm's length, describing the issue as 'an internal Ukrainian affair'. The Ukrainian government has recognised that the Crimean problem cannot be resolved by force. Crimeans overwhelmingly supported Kuchma for president, expecting that he would both support increased

autonomy and take steps to halt the ravaging of the Crimean economy, and his election in the summer relieved a lot of pressure. And Kuchma's son-in-law was appointed Prime Minister of Crimea in October, ensuring a strong link between the local Crimean authorities and Kiev. In early 1995, Meshkov's own incompetence finally secured his downfall. It had taken him six months after his election to nominate a government. To the irritation of Crimean parliamentarians, he then named non-Crimean Russians to important posts as well. By September 1994, the Crimean parliament had had enough, and stripped him of his powers. As the economy descended into chaos and politics into corruption, Crimeans were ready to tolerate even further censure of their once-popular president. In March 1995, the Ukrainian parliament annulled the Crimean constitution and abolished the post of Crimean presidency. It also recommended charging Meshkov with abuse of office. Scarcely a murmur of protest greeted the decision in Crimea.

Economic Reform

Kuchma inherited an economy in tatters. In 1993, inflation had reached 10,200%. According to various estimates, production declined by as much as 50% since 1990. The budget deficit stood at 20% of GDP, and the balance of payments deficit was over $3bn. Moreover, the government was riddled with corruption. On paper, Ukraine's economy was the worst of any country of the former Soviet Union that was not at war. But the books do not tell the full story: while officially the economy was in crisis, many Ukrainians were supplementing their state work with activities in the largely unofficial economy. According to some estimates, at least half of Ukraine's $6.5bn in exports in 1993 went unreported to the government. The state's domination of the official economy forced many to move into the unofficial economy in which a market was already at work, albeit one sometimes shaped by criminal gangs rather than market forces. This unofficial economy was waiting to be tapped by economic reform.

Kuchma campaigned on a platform of economic reform and intolerance of corruption. After an early period of consultation and delay, he presented his comprehensive reform plan to parliament on 11 October. After a week's debate, it was passed by a vote of 231 to 54. Kuchma's plan contains three basic elements:

- Financial stabilisation through deep cuts in subsidies, budgetary restraint, the introduction of a national currency, the rationalisation of the tax regime with the aim of increasing revenues, and eliminating stringent export controls to facilitate the flow of trade.

- Sweeping privatisation of state-owned enterprises in all sectors, except those which contain natural monopolies; privatisation of agri-

cultural land and distribution networks, as well as housing, to proceed at a slower rate.

• Price liberalisation of all goods, with gradual increases in prices
on such vital commodities as housing and energy.

Kuchma has begun to implement this plan through a series of presidential decrees that freed some prices and began a privatisation programme for approximately 8,000–9,000 medium- and large-sized industries. With international help, in the last quarter of 1994, Ukraine also
began to pay off its energy debt. Difficult issues lie ahead, however. The
state budget was under debate in the Rada and a second round of price
increases was scheduled for the first quarter of 1995. While the President's
administration and cabinet are dominated by young reformers, in spring
1995 some key positions were still held by anti-reformers. The prime
minister that Kuchma inherited from Kravchuk was only replaced by
Yevgenii Marchuk in March. In addition, Kuchma himself has preferred to
move cautiously, even as his programme faces long-term opposition from
the left and increasingly from the entrenched state economic bureaucracy.

Ukraine must supplement its internal reform efforts with financial
help from abroad. Estimates run in the range of $4–5bn over the next three
years, to be procured from international financial institutions, members of
the G-7, and from rescheduling Ukraine's energy debts by its main suppliers, Russia and Turkmenistan. Thus far, Ukraine has received a $371m
loan from the IMF, and smaller contributions from the governments of the
Netherlands, Germany, the United States and Canada. Whether Ukraine
can acquire other outside support will be crucial in sustaining internal
support for reform.

Ukraine must also address its looming energy crisis. Approximately
90% of its oil is imported from Russia, while 100% of its natural gas is
imported from Russia and Turkmenistan. This is supplemented by domestically produced nuclear energy which provides Ukraine with approximately 40% of its electricity. Ukraine cannot afford to pay Russia and
Turkmenistan for the oil and gas, and many of its nuclear reactors, including the one at Chernobyl, are unsafe. The government is exploring alternative energy sources, but for the foreseeable future Ukraine must rely on
current sources, thereby increasing Ukraine's indebtedness to Russia and
Turkmenistan. And continuing to use unsafe reactors may alienate European sources of assistance.

Relations with Russia

Ukraine's security policy is shaped by the need to construct a long-term
relationship with Russia. The large ethnic Russian population within
Ukraine itself means that Ukraine's policy towards Russia has important

domestic ramifications as well. The pattern of Ukrainian–Russian relations has not been a simple one. Yet the relationship has been more cordial and pragmatic than the rhetoric has suggested or than many in the West feared. Both sides understand the need to avoid crisis; Ukraine from early on, and Russia – gaining an appreciation of the potential costs of Ukrainian instability – from late 1993 and early 1994 onwards.

But the main elements of dispute that existed as 1994 opened remain in early 1995: energy supplies, debt repayment, and Ukraine's relationship to the CIS. In particular, they remained divided over the Black Sea Fleet. There have been a number of 'resolutions' of this problem at the highest level since 1992. All have foundered in follow-on technical talks. The main stumbling block is not how to divide the Fleet, which has been more or less agreed for some time now, but the terms for basing the Russian component in Crimea. Ukraine has resisted proposals that, in its view, shift *de facto* sovereignty of the bases to Russia.

In addition, basic tensions remain in the relationship that prevent issues from being resolved. For Russia, permanent resolution of these issues risks admitting once and for all that Ukraine is a state within its current borders. For the foreseeable future, all but the most extreme Russian nationalists understand that Russia's current internal crisis is best dealt with if Ukraine is standing on its own feet rather than serving as a drain on Russia. But these same analysts and policy-makers hope for a long-term evolution of Ukrainian independence in a manner that deepens integration of the two states. No one in Moscow wants to reach a permanent settlement that forecloses this option.

Ukraine's long-term preoccupation is with Russia's potential strength, for Russia will inevitably emerge as the stronger of the two. Bilateral negotiations reflect this relative distribution of power. Ukraine can resist Russian demands, but cannot obtain the permanent settlement it desires. This places the current relationship in a limbo of indeterminable length, in which both sides find *ad hoc* pragmatism and cooperation in their interest, although neither appears willing or capable of placing the relationship on a more solid, more enduring footing.

One interesting trend that may help to resolve this dilemma is the increasing multilateralisation of at least some formerly bilateral issues. A trilateral mechanism dealt with nuclear matters, and the G-7 and international financial institutions are now turning bilateral economic issues, such as the energy debt, into multilateral assistance plans. Similar approaches may well help to resolve outstanding bilateral disputes.

The Challenges Ahead

Ukraine faces enormous challenges in the months ahead. Its economy is still in decline, with fundamental questions of the legal and financial framework for market reforms still unanswered. Politically, although the opposition to reforms is in disarray, President Kuchma must sustain his political momentum. Hesitation or retreat on his part would be enough to reunite them. Kuchma faces real tests of his political acumen as he works to win adoption of the law on power, refashion the structure and personnel of the executive branch and sustain his economic and political reforms. A key challenge will be to hold off the ambitions of his own prime minister, Yevgenii Marchuk, a tough professional from the internal security forces, who is widely seen as an effective leader and a possible alternative to Kuchma should his reforms falter.

In the past year, many outside observers have compared Ukraine to a state on the edge of collapse. For Ukraine, however, the comparison that matters most is not with abstract models or outside expectations of performance, but with its own history. The current situation, despite its problems, remains the best chance Ukraine has ever had – or is likely to have – for the emergence of an independent Ukrainian state. The brief opportunity that arose earlier this century was strangled by war and civil conflict. In the near term, Russia's own internal challenges have opened up a breathing space for the consolidation of Ukrainian independence. The thread that runs through most of the important decisions made last year by Ukrainian leaders and citizens alike is the realisation that while this breathing space exists now it will not last forever, and that the greatest immediate danger facing Ukraine is not foul play, but suicide.

The Balkan Battlefields

The relief of Sarajevo, the end of hostilities between the Bosnian Muslim and Bosnian Croat forces, and the cease-fire agreement signed by Krajina Serbs and Croat authorities raised expectations in the early part of 1994, however faint, that an end to the wars of the Yugoslav succession might be in sight. At the very least, it was hoped that the external powers would succeed in coordinating their policies more effectively and thus be able to goad the parties in conflict towards resolving their outstanding differences.

Those hopes have been dashed. The course of events in the latter part of the year and in early 1995 demonstrated that the underlying sources of conflict remain as obdurate as ever. Indeed, developments in Croatia towards the end of 1994 raised the prospect of a return to all-out war between Croatia and Serbia. This prospect only temporarily receded with the decision in March 1995 by Croatian President Franjo Tudjman to allow the presence of a much smaller UN force with a changed mandate. Even in Macedonia, hitherto unscathed by the ravages of ethnic warfare, ominous signs of inter-ethnic tension threatened the wider stability of the region.

Lower-Level Warfare in Bosnia-Herzegovina

There was a significant reduction in the intensity of warfare throughout the Republic of Bosnia-Herzegovina in the first half of 1994. This was the result of two developments in February: the establishment of a heavy-weapon exclusion zone around Sarajevo, and the agreement reached between the Bosnian Government Army (BiH) and Bosnian Croat Army (HVO) on a general cease-fire. These developments enabled the humanitarian relief operations of the UN High Commissioner for Refugees (UNHCR), supported since September 1992 by the Bosnia-Herzegovina Command (BHC) of the UN Protection Force (UNPROFOR), to be carried out far more effectively than before.

Political progress towards an overall peace settlement for Bosnia, however, remained as elusive as ever, despite many efforts to inject some momentum into the stalled negotiations. Indeed, in the latter half of the year, the lack of progress and the growing evidence of divisions among the external powers over how to handle the conflict set the scene for a gradual deterioration of conditions throughout the republic. Although former US President Jimmy Carter was able to negotiate a four-month cease-fire and cessation-of-hostilities agreement in late December 1994, the onset of spring, as in previous years, has brought with it the prospect of renewed, and this time bloodier, fighting.

The Sarajevo Massacre and its Aftermath

On 5 February 1994, a single mortar shell exploded in a crowded market square in Sarajevo killing more than 60 and injuring nearly 200 bystanders. It was assumed that the shell had been fired by the Bosnian Serb Army (BSA) which was besieging the city, but formal investigations were never able to demonstrate this conclusively. Still, the resulting carnage and public revulsion led to irresistible pressure for action to be taken against the BSA. On 7 February, UK Foreign Secretary Douglas Hurd stated that the European Union must now seek to 'bring about the immediate lifting of the siege of Sarajevo by all means necessary, including air power'.

The previous day, UN Secretary-General Boutros Boutros-Ghali had already requested that NATO be prepared to launch air strikes against artillery positions in or around Sarajevo responsible for attacks on civilian targets when asked to do so by the UN. Responding to the request, the North Atlantic Council issued a ten-day ultimatum on 9 February calling on the Serbs to remove, or place under UN control, all heavy weapons within a 20km exclusion zone around Sarajevo; failure to do so would lead to air strikes.

The BSA did not begin to move out of the exclusion zone until 17 February, after the Bosnian Serb leader, Radovan Karadzic, had come to an agreement with Russia's special envoy, Vitaly Churkin. As NATO and

UN officials attempted to clarify command-and-control issues pertaining to the use of air power, Churkin had, in separate talks in Pale (the Bosnian Serb 'capital'), agreed to replace the Serbs in the exclusion zone with some 800 Russian peacekeepers. This unilateral Russian intervention removed the need for NATO air strikes, and thus was welcomed, but it was also a reminder of Russia's determination not to be sidelined by Western initiatives and its willingness to take unilateral action to assure that end. The creation of the exclusion zone around Sarajevo did, however, ensure that a semblance of normality was restored to the capital after a period of more than 22 months of daily bombardment and sniping by Serbian forces surrounding the city.

While Sarajevo entered a period of relative calm, the vulnerability of other so-called 'safe areas' continued to be cruelly exposed. In late March and early April, Serb forces surrounding the enclave of Gorazde in eastern Bosnia stepped up their bombardment of the city. In spite of repeated threats of NATO air action, Bosnian Serb forces under General Ratko Mladic continued the offensive against Gorazde, threatening to overrun it completely. Fearing that the enclave was about to fall, the UN Force Commander in Bosnia, Lieutenant-General Sir Michael Rose decided on 10 April to call for NATO air strikes against Serbian ground targets. The resulting air action, the first of its kind in NATO's history, did not put an immediate end to the offensive. Ignoring Russian President Boris Yeltsin's expression of anger at not having been consulted about the NATO operation, the North Atlantic Council issued another 'Sarajevo-style' ultimatum to the Serbs. This called for an immediate cessation of hostilities, the withdrawal of all forces from the town centre and the removal of heavy weapons from a 20km exclusion zone around the city. The Serbs responded by gradually decreasing their attacks on the city.

The 'Gorazde incident' led to a storm of misdirected criticism of the UN Secretariat's inability to protect the 'safe areas' which the Security Council in 1993 had designated under Resolution 836. The blame should have been directed at the countries that make up the Security Council. When the 'safe areas' concept was first discussed, the UN Secretariat, after consulting UN commanders in the field, outlined clearly to the Security Council the political and practical problems posed by their creation. In particular, it was stressed that in the continued absence of a political settlement, the viability of the 'safe areas' would depend crucially on a credible UN military presence in the designated areas. The then Force Commander estimated that it would require some 32,000 men to provide such a presence. Having voted to establish the 'safe areas', however, the UNSC members neither provided sufficient troops of their own, nor money to assure that troops from other nations could be put in place, to make the concept viable. When the Gorazde incident took place, even the

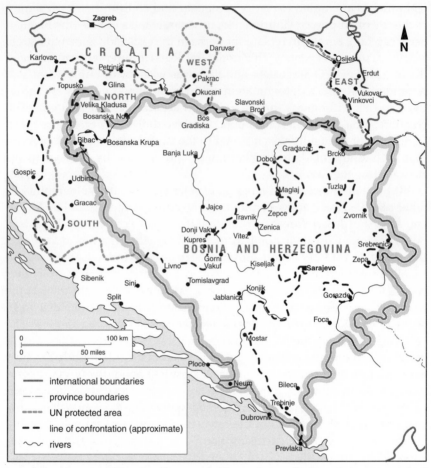

troop levels agreed by the Secretary-General as a 'light minimum option' (7,200) were nowhere near being met.

The Pattern of Military Operations in Bosnia

The Gorazde crisis was precipitated by a Serbian offensive against the enclave, but for the most part the Bosnian Serb forces have concentrated on consolidating gains made earlier in the war. Particularly in the latter half of 1994, it was the BiH, now far better equipped and disciplined than when it faced the initial Serb onslaught in 1992, that was on the offensive. Indeed, at one point in late September, UNPROFOR threatened air strikes against Bosnian government forces around Sarajevo if they continued to violate the cease-fire.

Throughout 1994, the BiH underwent a painful learning process, during which it gradually abandoned previously used crude and ill-considered tactics. For too long its basic strategy was to concentrate its superior manpower and attack Serbian forces at particular points along the confrontation line. This, however, played directly into the hands of the Bosnian Serbs who have always been able to concentrate their superior fire-power at the points of enemy pressure. In July 1994, BiH strategy had disastrous results around Mount Ozren where government forces were lured into a direct engagement with Serbian forces and as a result suffered 2,500 casualties without making any significant gains.

In the wake of this operation, the Bosnian government forces changed strategy. Rather than large operations, they launched several smaller simultaneous offensives at various points along the confrontation line, relying on interior lines of communication and superior manpower to put pressure on the Serbs. Improved training and discipline also began to yield results on the battlefield. In August, the 5th Corps of the BiH overran Bihac which for some time had been controlled by Serb-backed rebel Muslim forces under the leadership of Fikret Abdic, a Muslim entrepreneur turned warlord. Following this success, BiH operations intensified throughout the country. By mid-October, BiH forces were active around Bugojno in central Bosnia; the Maglaj finger; the Doboj finger around Gradacac; as well as around Visoko and Olovo. In early November, in its first major gain since the war started in April 1992, BiH forces supported by HVO troops captured the city of Kupres.

Yet the limits of what the Bosnian government can achieve by purely military means were made clear during the Bihac crisis. In late October and early November the BiH, still exultant at its victory over Abdic's rebel forces in August, launched an offensive out of the enclave. For about two weeks, the much-vaunted 5th Corps made rapid progress, overrunning Bosanska Krupa and capturing large quantities of Serbian equipment. On 9 November, however, government forces were faced with a large-scale Serb counter-attack, involving tanks, artillery and air power, coordinated and supported by Krajina Serbs in Croatia as well as by remnants of the Muslim forces loyal to Abdic. Within less than a week, most of the territory the government had gained had been recaptured and Bihac itself came under sustained artillery shelling and air attack from Croat Serb bases in Krajina. Yet again, the Serb forces appeared to be on the verge of capturing a UN-designated 'safe area'. Responding to this threat, NATO aircraft went into action, attacking the Udbina airfield in the self-styled Republic of Serbian Krajina (RSK) on 29 November and, a few days later, bombing Serb surface-to-air missile sites in the RSK and near Bosanska Krupa.

The pattern of military confrontation inside Bosnia over the past 16 months shows that, although the arms embargo has not prevented significant quantities of equipment from reaching either Croatian or Bosnian forces (the Croatian Air Force, for example, has increased the number of its operational MiG-21s from 3 to 21 while the embargo has been in place), the overall military balance is relatively unchanged. The Bosnian Army is superior in numbers. Bosnian Serbs, while fewer in number, are still better equipped and, above all, better trained. Without a considerable increase in the amount of outside assistance no military 'victory' can be achieved by either party.

The Contact Group Initiative

It was partly against this background that renewed attempts were made by outside powers to push the diplomatic process forward and to work 'as a matter of urgency towards the full cessation of hostilities'. The Contact Group (France, Russia, the UK, Germany and the US) was formed on 26 April 1994 in an attempt to inject new momentum into the peace process. The Group developed a new set of peace proposals which were presented in map form in early July. The map assigns 51% of the Republic's territory to the newly formed Bosnian-Croat Federation and 49% to the Bosnian Serbs. Its unveiling was accompanied by a warning to the self-styled Bosnian Serb government in Pale that failure to accept it would lead to 'total isolation' and a tightening of sanctions against its main ally, the Federal Republic of Yugoslavia (FRY).

The Bosnian government described the plan as 'seriously flawed' because it allowed the Serbs to hold on to 'ethnically cleansed' territory, but it approved the plan under pressure on 17 July. The authorities in Pale, however, effectively rejected the plan on 19 July by insisting on further talks. Their objections continue to block progress in negotiations and make eventual agreement doubtful. In the first place, they demanded that the enclaves of Gorazde, Zepa and Srebrenica be under Serb control, while Sarajevo should be divided into two ethnic halves. Second, Pale insisted that the strategically vital Posovina corridor linking Serbia proper to Serb territories in the north around Brcko must be widened. The Bosnian Serbs also argued that they should be allowed to enter into a confederation with the FRY.

In an effort to move things forward the Contact Group tried to use the increasingly strained relations between President Slobodan Milosevic in Belgrade and the authorities in Pale as a lever to pressure the Bosnian Serbs into accepting the plan. Relations between Radovan Karadzic and President Milosevic were already extremely poor. Now that Milosevic was being threatened by further and tighter sanctions, he came out in favour of the Contact Group plan and denounced its rejection by the Bosnian Serbs as 'senseless and absurd'. Milosevic accused his former allies of being driven not by the greater interests of Serbdom, but by 'mad political ambitions and greed'.

In early August Milosevic closed the border between Serbia and Bosnia both to increase the pressure on the Bosnian Serbs and to demonstrate his basic acceptance of the Contact Group proposal. Although the border was not completely sealed, the cut-off has threatened the fuel supply on which the Bosnian Serb Army is dependent, and has further

The Contact Group Map as of 1 January 1995

soured relations between Belgrade and Pale. More importantly from Milosevic's point of view, it has enabled the FRY to begin to repair relations with the Western powers. Thus, in response to Milosevic's decision to allow 135 monitors to be deployed along Serbia's border with Bosnia, the Security Council, on 24 September, approved a resolution lifting a limited number of sanctions on the FRY (those relating to civilian air traffic, cultural and sporting links).

Bad Blood Between the Alliance and the Contact Group

The decision to cultivate relations with Milosevic as a means of forcing the Bosnian Serbs into accepting new peace proposals has also exposed the fragile unity of the Contact Group. Although the Group has sought to project an image of unity and resoluteness, tensions among its members have been evident from the outset. They became increasingly obvious, however, after the Bosnian Serbs rejected the plan in July and the deteriorating situation in the country until former US President Carter brokered a cease-fire in December.

At the heart of these tensions is the basic difference in perception of the origins and nature of the conflict in the former Yugoslavia held by France, the UK and Russia on the one hand, and the US on the other, with Germany tacitly supporting the US but anxious not to upset its European allies. When the Republican leader of the US Senate, Robert Dole, told the UN in November 1994 'to get off NATO's back and let NATO take care of Serbian aggression', he was voicing a much deeper division among the external powers about the perceived root causes of the conflict.

For the most part, opinion in the US has been sceptical that divisions among the Serbs could be exploited and has tended to view Serbdom as an undifferentiated whole, controlled and manipulated entirely from Belgrade. This general disposition to see the conflict primarily in terms of Serbian aggression has been reflected in the early position taken by the administration, which has been continued even more strongly by Congress, for a unilateral lifting of the arms embargo against Bosnia. Indeed, on 11 August President Bill Clinton laid down a deadline (15 October) for the Bosnian Serbs to accept the Contact Group plan; if not the US would go to the Security Council and ask for the embargo against Bosnia to be lifted. While none of the other Contact Group members deny the culpability and central role of Serbia in the Yugoslav conflict, the threat of a unilateral and selective lifting of the arms embargo has complicated the work of the Contact Group. The Group managed to convince the US not to press for adherence to its October deadline, and in 1995 the pressure to lift the embargo came more from Congress than from the administration.

A further, and closely related, source of tension between the US and other members of the Group has been the greater readiness of the US to use air power against the Serbs in situations where Britain, France and Russia all felt that UNPROFOR's role on the ground would be unduly compromised. On several occasions, not only Russia but also NATO members contributing troops to UNPROFOR have sought to impress on the Americans that the UN has not in fact been issued with a war-fighting mandate. Excessive use of force by NATO would only destabilise the operation and draw the alliance into the conflict. The French, less inhibited about rupturing the illusion of allied unity, have repeatedly deplored US 'irresponsibility', as when French Foreign Minister Alan Juppé, in early December, angrily referred to those who 'teach us lessons daily and have not lifted a little finger to put even one man on the ground'.

NATO officials have been highly critical of the 'dual key' arrangement eventually agreed between the UN and NATO as a way to manage the problem of who decides when to bomb in the former Yugoslavia, and this has led to much talk about UN–NATO tensions. But just as serious a rift has been running through the alliance itself. Although the US insists it is not withdrawing from *Operation Sharp Guard*, the NATO operation enforcing the arms embargo in the Adriatic, its unilateral decision no longer to stop ships headed for Bosnia was both an indication of this tension and contributed to it, even though it did not much affect the NATO operation itself.

The Contact Group suffers from other disabilities. While Russia's position has often tended to converge with that of the UK and France, it has been anxious to ensure that its own interests are not ignored by the NATO members of the Group. The role it played during the period of the Sarajevo ultimatum and its open criticism of the NATO air-strike policy later in the year highlighted Russia's concern. In spite of the evident strains among its members, the Contact Group continues to work on the basis of its agreed peace proposals, though progress is slow and threatens to be undermined by developments elsewhere, most notably in Croatia.

The original plan presented in July did not clarify the constitutional relationship between the part of Bosnia set aside for the Bosnian Serbs and Serbia proper. To secure eventual support from the Bosnian Serbs, the Contact Group foreign ministers agreed in December on a text that alluded to a possible confederation between Bosnian Serbs and the FRY within the framework of the Contact Group peace plan. Although Milosevic warmly welcomed this as a 'major step forward', Radovan Karadzic remained opposed. Even if Belgrade and Pale were to agree with a Contact Group formula, the entire plan still rested upon the increasingly questionable assumption that the Bosnian-Croat Federation, which would control 51% of the country, is a viable entity.

The Washington Agreement and the Bosnian-Croat Federation
Public coverage and analysis of the war in the former Yugoslavia tends to ignore the role Croats have played in the destruction of Bosnia-Herzegovina. This is a curious case of neglect given that much of the bloodiest fighting in the republic took place between Muslims and Croats in western Herzegovina following the collapse of the Croat–Muslim alliance in April 1993. Moreover, in his attempt to carve out a separate Bosnian Croat state, the Bosnian Croat leader Mate Boban was fully supported by Croatian President Tudjman. Regular Croat Army (HV) units fought alongside, and coordinated their activities with, the HVO inside Bosnia. That this constituted a case of international aggression identical to what the Serbs were accused of elsewhere in Bosnia was for a long time simply ignored by the outside world, especially by Germany and the US, the countries on which Croatia relied for support. In addition, HVO forces were responsible for some of the most horrific war crimes committed in the whole of the former Yugoslavia (notably in its actions in their Medac pocket). It was only in late January 1994 that the US began to raise the question of imposing international sanctions against Croatia.

By then the US and Germany were also pressing hard for the Croats and Muslims to reach a cease-fire. In this they succeeded and a general cease-fire agreement signed on 24 February 1994 between the HVO and the BiH brought to a temporary end one of the most destructive phases thus far of the wars of the Yugoslav succession. US mediation and support was also critical for the formation of a federation, whose territory covers the areas in which Bosnian Muslims and Croats constituted the pre-war majority. The agreement on the creation of such a federation, and 'preliminary agreement' on the establishment of a confederation between the Bosnian Republic and Croatia, were signed in Washington in mid-March.

The detailed Federation Constitution, which envisaged a powerful federal government responsible for defence, foreign policy and economic affairs, was signed in May. The Federation would be based on a system of power-sharing between the two ethnic groups, including the annual rotation of the offices of president and prime minister. At the regional level, eight Swiss-style cantonal governments would be established providing the basis for the reintegration of ethnic communities split by the recent war.

Actual progress in establishing closer relations under the new Federation, however, has been extremely slow. The EU-administered city of Mostar, the scene of the most intense fighting in 1993, is still divided along ethnic lines and in November 1994 'extremist' Croats were allegedly still sniping at Muslim civilians in the city. In September, Bosnian radio claimed that joint Croat–Serb attacks had been launched against Bosnian positions in Bihac, and the following month local Muslim officials in the Neretva canton alleged that rather than working to reintegrate ethnic communities, Croats were continuing with policies of 'ethnic cleansing'. Indeed, these events, as well as President Tudjman's *volte face* from a position strongly supporting Mate Boban's ambitions to set up a separatist Croat Republic of Herceg-Bosna to one in which he welcomed the Federation agreement as 'historic', suggest that the description of the Federation as little more than a 'glorified cease-fire' is not altogether inaccurate. It is unlikely that the federation will survive once it is deemed by Tudjman and Bosnian President Alija Izetbegovic to have outlived its tactical usefulness.

The record of military cooperation between the Croats and Bosnian forces, for which the Federation agreement makes provision, also argues that the agreement is less than it seems. Even though the US has been pressing hard for the creation of an effective joint command (in November the US sent General John Galvin, former Supreme Allied Commander Europe (SACEUR), to Zagreb to urge both parties to cooperate), limited progress has been made. The capture of Kupres in early November by joint Croat–Bosnian action is very much the exception.

The US administration, whose diplomacy brought the Federation into being, remains strongly committed to its survival. There can be no doubt that its existence, however shaky, has prevented a resumption of fighting between the BiH and HVO/HV. The future of the Federation, however, was more likely to be determined by developments elsewhere, above all in Croatia where the underlying Serb–Croat conflict was again dangerously close to the point of violent eruption.

Croatia: a Return to All-out War ?

Under the original UN plan for Croatia, organised by UN negotiator Cyrus Vance in January 1992, UNPROFOR was supposed to have overseen the disbanding or withdrawal of 'all military forces' inside three so-called UN protected areas (UNPAs), covering about one-third of Croatian territory. In these areas (Krajina, eastern and western Slavonia), Serbs constituted a majority of the population. Because none of the parties to the agreement, especially the Krajina Serbs, has cooperated, UNPROFOR, which was never given an enforcement mandate, has been unable to implement the original plan. For nearly three years the situation has been deadlocked. The Krajina Serbs continue to insist that the RSK deserves recognition alongside other entities in the former Yugoslavia, while Zagreb has called on the UN to end the Serbian 'occupation' of its territory. The situation in Croatia has remained relatively stable compared to the fighting in Bosnia since the failure of a major Croat military offensive against Serb positions in Krajina in January 1993.

On 30 March 1994, Vitaly Churkin, Boris Yeltsin's special envoy to the former Yugoslavia, succeeded in brokering a cease-fire agreement between representatives of the RSK and Croatian authorities. Under the terms of the agreement, all troops deployed on the front-line, some 1,000km from Drnis in southern Croatia to Vukovar in the east, would be withdrawn by at least 1km. Heavy weapons would be pulled back by 10km. The resulting buffer zone was to be monitored by UNPROFOR forces in Croatia. The cease-fire came into effect on 4 April and was an essential preliminary step to 'normalising' relations between the Krajina Serbs and authorities in Zagreb.

Although mutual suspicions have remained profound, surprising progress in attempts to improve relations was made towards the end of the year. Under the auspices of the International Conference on the Former Yugoslavia (ICFY), the Croatian government and the RSK signed an economic agreement on 2 December 1994 which called for the re-establishment of key services (water, oil and electricity supplies), as well as the opening of transport links between the two parts of the country. By mid-January, the main motorway running eastwards from Zagreb through Serb-controlled Sector West had been reopened to traffic. A critically important oil pipeline for the Croatian economy, connecting the Adriatic coast to Central Europe, has also been opened under the agreement. Although President Tudjman made it clear that full reintegration remained a non-negotiable objective, this limited *rapprochement* was significant given the centrality of the Serb–Croat conflict to the whole region.

In January 1995, however, President Tudjman reversed the process by his decision to terminate the UN operation in Croatia. If the UN did not leave, he argued, Croatia would soon become another Cyprus where the UN presence had frozen a *de facto* division of the island with no political solution in sight. Citing the inability of the UN to deliver on its mandate and to ensure the 'reintegration of the occupied territories into Croatia', he informed the Secretary-General that the UNPROFOR mandate would be terminated by the end of March.

President Tudjman's decision was partly linked to his own internal position in Croatia. Having repeatedly threatened not to renew the mandate he had painted himself into a corner in relation to those in his own party and within the military who favoured more forceful action to retake the Serb-controlled lands. Tudjman also appears to have believed that there was a credible available military option. Leading officers within the Croatian military, who have spent the last three years re-arming and restructuring their armed forces while watching the Krajina Serbs become increasingly isolated, appeared to share Tudjman's assessment of the changing military balance in the region. And the strains that had developed between Knin and Belgrade may have led President Tudjman to the dangerously erroneous conclusion that Milosevic would refrain from intervening in a renewed war between Croatia and the RSK.

There is in fact precious little evidence to suggest that the Yugoslav Army would remain passive if a large-scale attack was launched against Krajina. Moreover, the Krajina Serbs could count on the support of fellow Serbs in neighbouring Bosnia, whom they have recently assisted, and continue to assist, with extensive military force in operations against the Bihac enclave. The commander of the Bosnian Serbs, General Mladic, was himself based in Knin before he took over command of the BSA.

The prospect of a return to all-out warfare between Croatia and Serbia has forced the outside powers to coordinate their policies. Under intense pressure from the US and Germany, Tudjman in early March 1995 agreed in principle to an extension of the UN presence, but at a much lower level. In the negotiations for a new mandate (in which the US played a direct role), a possible reduction of UNPROFOR from roughly 12,000 to somewhere between 5,000 and 8,000 was discussed. The task of this force would be to control the Zone of Separation, to monitor the international border between Croatia and Bosnia, and to help implement the economic plan of the Z4, the four-member (the US ambassador in Zagreb, his Russian counterpart, and two representatives of the ICFY) contact group for Croatia.

These kinds of compromises, however, would only buy time, and did not address the underlying grievances of the parties. Moreover, the Krajina Serb position with respect to a new mandate was not taken as much into account as was President Tudjman's position. Although the RSK's formal approval was not needed, if it was not prepared to cooperate, as it was not with the original Vance plan, the UN's already thankless task would become even more hopeless.

The Threat of Continued Instability

The basic objectives in the former Yugoslavia on which the outside world has been able to agree have been threefold: lending 'good offices' and encouraging the parties to reach a negotiated solution; relieving the humanitarian consequences of the war in Bosnia-Herzegovina; and preventing the conflict from spreading. These aims have been reflected in specific tasks given to UNPROFOR in Croatia, Bosnia and Macedonia. At no stage has the UN been asked, let alone given the resources, to engage in coercive action in an effort to impose a solution on the warring parties.

In part, these minimal objectives reflect a basic lack of consensus about the origins and nature of the conflict; a lack clearly proclaimed by the divisions among the members of the Contact Group. The espousal, and uneven application by the 'international community', in the post-communist and multi-ethnic context of the former Yugoslavia of the principles of the right to self-determination and the inviolability of borders have led to suffering on a scale not seen in Europe since the Second World War. Until the outside powers coordinate their policies with a view to mitigating the consequences of applying these competing principles, they will be unable fundamentally to affect the wars of the Yugoslav succession. Meanwhile, the bitter antagonists in the conflict will continue to play on and exploit differences between the powers, something at which they have become increasingly adept.

Although it has made little progress in the promotion of a negotiated solution, the UN has been rather more successful in carrying out its other two basic objectives. Yet here too recent events throw a dark shadow over future developments. The humanitarian effort has been impressive. The prospect of an intensification of the war after the cease-fire ends in April will again threaten UNHCR activities in Bosnia.

Efforts to ensure that the conflict does not spread southwards have focused on the former Yugoslav republic of Macedonia. In 1993 the UN deployed a peacekeeping force, which included US troops, as a shield against any aggressive intentions. It has generally been regarded as one of the more successful aspects of the UN's involvement in the former Yugoslavia. Even here there are ominous signs. In February 1995, Macedonian police clashed violently with ethnic Albanians in the western town of Tetovo, leaving one demonstrator dead and more than 20 others injured. The incident was sparked by the creation of a university for ethnic Albanians in defiance of a government ban. If this should be the precursor to a general breakdown of ethnic relations in Macedonia, the dangers of the conflict in Yugoslavia expanding in a new and disturbing dimension, and bringing in neighbouring states, will have been seriously increased.

A Modest Recovery in the West

In 1994, Western Europe managed to climb out of the trough of despondency that had characterised much of the previous year. The Maastricht Treaty on European Union, which entered into force on 1 November 1993, began to have an impact, negotiations on EU enlargement were successfully concluded, a pre-accession strategy was agreed for six Central and East European countries, economic growth began to pick up and there was a growing consensus around Commission President Jacques Delors' White Paper on 'Employment, Growth and Competitiveness'. At the same time, however, there was little progress in ending the conflict in Bosnia and a damaging rift appeared in transatlantic relations concerning the possible lifting of the arms embargo. NATO became embroiled in a dispute with the UN over policy as well as command and control of forces in Bosnia.

The Union's Constituent Parts

There were a number of elections in 1994; in the two most important, Helmut Kohl won an unprecedented fourth term as German Chancellor, while Silvio Berlusconi swept to power in Italy only for his coalition government to fall apart by the end of the year. The left returned to power in Sweden and held on in Denmark. In the Netherlands an inconclusive result led to a coalition of left and right. In France, President François Mitterrand was increasingly isolated from the centre-right government of Edouard Balladur, while in the UK the Conservative Party's fortunes continued to decline.

In Germany the SPD under its new leader Rudolph Scharping had high hopes of winning the general election. A stream of regional election victories had given the SPD a solid majority in the Bundesrat and polls at the beginning of the year showed them ahead of Kohl's CDU. But gradually Kohl re-established his popularity, helped by a timely economic recovery, most noticeably in the east. There had been some speculation that the FDP might improve its waning electoral fortunes by switching horses, but the election of a new president, Rainer Herzog, in May revealed that the FDP was staying loyal to its coalition partners. In the October general election the FDP just managed to stay in the Bundestag, but its position was substantially weakened.

The SPD, Greens and PDS (the former East German communists) all managed to increase their representation, but just failed to oust the CDU/CSU/FDP coalition. Kohl's overall majority fell to only ten seats which led to speculation as to whether his coalition could survive a full four-year term. Foreign and security policy had played a minor role in the election campaign, but Kohl's victory confirmed in office Europe's strongest advocate of a federal EU. The CDU-led coalition was also more favourable than the SPD to the concept of sending German forces overseas on peacekeeping missions.

In Italy a sea-change in politics occurred with the demise of the traditional parties of left and right, nearly all tainted by corruption. Media tycoon Silvio Berlusconi's new movement *Forza Italia* dominated the March 1994 election campaign by promising deregulation, tax cuts, fiscal decentralisation and a reduction in the public debt. An alliance led by *Forza Italia* and including the Northern League and the neo-fascist National Alliance won a majority of seats in the lower house of parliament. The new government was unstable from the beginning and proved incapable of tackling major issues such as reducing public-sector debt. As the year progressed the familiar smell of corruption came to engulf Berlusconi who was forced to resign in December. On the foreign-policy front, the new government pursued a more nationalist approach – it blocked the EU opening negotiations with Slovenia on association because of a bilateral dispute over the expropriated property of Italians in Slovenia.

In the UK, despite modest economic recovery, the government's popularity continued to fall, partly as a result of numerous self-inflicted scandals and partly because of continuing deep divisions over Europe. In November the whip was withdrawn from nine Conservative rebels who voted against increasing the UK's contribution to the EU budget. This move effectively meant that the Conservatives no longer commanded an overall majority in parliament. The Labour Party elected Tony Blair, a proponent of 'modernisation', as leader to succeed John Smith who died suddenly in May. Blair's popularity with all sections of the population led to a commanding lead in the polls for Labour. He began moves to drop 'Clause Four', the Party's traditional commitment to public ownership, from the constitution and to adopt a far more positive approach towards the EU including foreign and security policy integration.

In France, President Mitterrand approached his last year in office in a state of ill health which prompted endless speculation about his succession. The withdrawal of Jacques Delors from the presidential race in December 1994 seemed to signal an easy victory for Edouard Balladur. The prime minister had a good year despite a number of corruption scandals. His handling of the hijacking of an Air France aircraft in Algeria in December further heightened his popularity. But in February 1995, a wire-tapping scandal caused a sharp reverse in his fortunes, his arch-rival Jacques Chirac stormed to a lead in the polls, and the choice of Lionel Jospin as Socialist candidate tightened the race again.

Politics and Economics in the European Union

Franco-German cooperation in EU affairs continued as Bonn and Paris agreed to coordinate their presidencies which spanned the second half of

1994 and the first half of 1995. There were, however, scarcely concealed differences of opinion on the future course of European integration and the priority which the EU should accord to the east as opposed to the south. A compromise formula was agreed at the Essen summit in December which established major programmes for the East Europeans and the Mediterranean countries.

At the European Council summit in Corfu in June, UK Prime Minister John Major, under pressure from the Eurosceptic wing of his party, vetoed the appointment of Jean-Luc Dehaene, the Belgian Prime Minister, as president of the Commission. The British objected to the way in which Dehaene had been selected by France and Germany, acting in collusion and behind the scenes. A special summit was called three weeks later to confirm Jacques Santer, the Luxembourg Prime Minister, as Delors' successor.

Santer's first difficult task was to arrange an agreed distribution of portfolios for his expanded 20-member Commission. Inevitably his choices were criticised, perhaps most heavily by those, like Sir Leon Brittan, who did not get the positions they hoped for. Santer retained overall control of the Common Foreign and Security Policy (CFSP) himself and divided external relations into four portfolios. Hans van den Broek became responsible for Eastern Europe, Russia and the former Soviet Union, Sir Leon Brittan for trade and industrialised countries, Manuel Marin for the Mediterranean, Middle East and Latin America and Joao Pinheiro for the developing countries. This geographical split replaced the functional political–economic split which had not proved successful in the last Commission.

Socialists dominated the new European Parliament which was elected in June. Klaus Haensch was elected president of a parliament determined to build on the increased powers it had gained under the Maastricht Treaty. It flexed its muscles almost immediately over the distribution by Santer of tasks among the Commissioners, forcing him to justify his choices and in some cases to make adjustments to take its view into consideration.

On the economic front, EU members began at last to move out of recession and even achieved a respectable average growth of 2.4% The second stage of Economic and Monetary Union (EMU) came into effect in January 1994 with the establishment in Frankfurt of the European Monetary Institute as a precursor to a European Central Bank. Renewed economic growth in 1994 increased the prospects for meeting the criteria and timetable for establishing an EMU and a single currency, as set out in the Maastricht Treaty. Under the Treaty, if a majority of member-states meet the criteria, a single currency could be set up between them in 1997; at the latest, a single currency would be established in 1999 between those member-states that met the criteria.

But instability in the financial markets in early 1995 – caused in part by the collapse of the UK bank Barings and the continuing depreciation of the US dollar – dashed EMU hopes again. The EU's Exchange Rate Mechanism, which in August 1993 had had to be altered to allow currencies to fluctuate within a 15% band rather than a 2.25% one, was put under pressure again in March 1995, forcing the devaluation of the Spanish peseta and the Portuguese escudo. High unemployment, which topped 20% in Spain and averaged 9.5% in the EU, also added to the difficulties of reaching the EMU criteria. Undeterred, French Prime Minister Balladur and Commission President Santer continued to argue that a single currency would be set up in 1997, although meeting that deadline seemed improbable.

Towards a Wider Europe

During 1994 the EU completed enlargement negotiations with four European Free Trade Association (EFTA) countries, Austria, Sweden, Finland and Norway. Two main reasons spurred these countries to submit membership applications. The first was the obvious success of the single European market programme and the recognition that only full EU membership would allow them to participate in decisions that would affect them even if they remained outside the EU. The second was the end of the Cold War. Previously, the political elites in the neutral states (Austria, Finland and Sweden) had acknowledged that their foreign-policy status was incompatible with EC membership. But the geopolitical changes which hit Eastern Europe in 1989–90 also changed this. Even those most wedded to neutrality swiftly accepted that this would no longer be a bar to full membership despite the fact that the Maastricht Treaty envisages an eventual common defence policy 'which might in time lead to a common defence'.

During the negotiations, specific problems arose over fisheries and energy (Norway), transit (Austria) and alcohol monopolies (the Nordics), but agriculture was far and away the most difficult hurdle because it was the economic sector requiring the biggest adjustments by the four applicants. The EU insisted that, because of the Single Market and the need to avoid establishing border controls for goods traded between member-states, the acceding countries would have to adopt the EU's lower agricultural prices. The applicants finally accepted this; to sweeten the deal, an agreement was reached on the payment of national subsidies over a five-year period to farmers in Austria, Finland and Norway, to compensate for the reduction in prices required by adjustment to the EU's Common Agricultural Policy (CAP).

Although Austria, which had applied in July 1989 (before the Berlin Wall fell), had entered a reservation on its neutrality in its application, a consensus was reached in the negotiations between the EU and the applicants on the CFSP. All sides agreed that when they became members the acceding states would be ready and able to participate fully and actively in the CFSP. The acceding states would accept without reservation all of the Maastricht Treaty's objectives and all of the provisions on the CFSP. The acceding states would also have to support the EU's specific policies in force at the time of the accession, meaning in particular the policy towards the former Yugoslavia. All the new member-states, for example, would have to participate fully in EU sanctions (which could be considered a violation of strict neutrality), and in discussions on policies with defence implications.

In making these commitments, the applicant states went further than Denmark had been willing to go. After the Danes had rejected the Maastricht Treaty in a referendum in June 1992, the Danish government negotiated several opt-outs; most crucially, Denmark would not partici-pate in EU decisions and actions that have defence implications. The insistence that new members accept the entire Maastricht Treaty can thus be seen as an attempt to cut short the trend towards 'variable geometry'. However, this does not affect Ireland, a neutral state that is already a member of the EU. The Maastricht Treaty's provisions on the CFSP state that it shall not prejudice the specific character of the security and defence policy of certain member-states, a reference to Ireland's neutrality.

Austria and Finland had already abrogated their agreements with Russia to remain non-aligned. None of the three neutral applicants, though, renounced their neutrality; because of the Irish precedent, the new member-states can argue that EU membership does not affect neutrality. But all three also tightened their cooperation with other European security organisations. In 1994, the three agreed to become observers of the WEU. All three have also agreed to participate in NATO's Partnership for Peace (PFP) programme, although Austria did not do so until February 1995.

Neutrality is no longer a central issue since the EU has yet to formulate a common foreign policy which has defence implications or to request the WEU to carry out such a policy. It could become a major problem, how-ever, if the 1996 IGC takes more drastic steps towards establishing a common EU defence policy.

Appropriate adjustments to the EU's institutions in the light of en-largement were made on the basis of the existing institutional provisions in the Maastricht Treaty – a decision which followed on from the conclu-sions of the Lisbon European Council in June 1992. This did not, however, prevent the UK from trying, through an adjustment to the EU's complex voting procedures, to preserve its ability to veto legislation by making it possible for the same minority to block action – even though there were more member-states. Apart from Spain, the British won no consistent support from other member-states and after holding out for a number of weeks were forced to back down on 27 March. Sweden and Austria were each granted four votes in the Council and Finland and Norway three. The qualified majority was fixed at 64 votes (62 after the Norwegian 'no') which maintained the existing share of votes necessary for a majority at around 71%. There would be a larger Commission, consisting of 21 Commissioners (later reduced to 20 when Norway dropped out), whilst the number of Members of the European Parliament was fixed at 22 for Sweden, 21 for Austria, 16 for Finland and 15 for Norway.

The European Parliament approved enlargement on 4 May, and the Accession Treaties were formally signed by the heads of state and government at the European Council summit on 24 June. Each of the applicant countries had agreed to ask for the consent of its people through a national referendum. Austria was the first on 12 June and secured a resounding 67% 'yes' vote; Finland followed on 16 October with a comfortable 57% positive vote; Sweden scraped through on 16 November with 52% supporting the negotiated deal; but on 28 November Norway, which had always been viewed as the most doubtful of the applicants, voted the proposition down with 52% against.

The speed with which negotiations were concluded showed the EU's willingness to accept the EFTA applicants at a time of some internal confusion, and the high degree of preparedness on the part of the applicant states. With the three new members the EU's territory will be a third larger, which will still make it less than half the surface area of the US and only a fifth as large as Russia. Its population will rise by 6.2% to just over 370m and its GDP by 7%. The enlarged EU will have a GDP nearly 30% higher than the US and twice that of Japan. It will also have acquired a 1,200km border with Russia.

The political attitudes of the new members and their negotiating strategy suggest that their accession will have a significant impact on a number of important policy areas. There can be no doubt that the newcomers will press for higher standards in environment and social policy. They are also likely to press for more openness and transparency in EU decision-making and to join those countries which demand that member-states enforce EU directives. They can be expected to be in the first wave of those countries able to meet the criteria for the EMU. Some newcomers are joining against a background of considerable Euroscepticism which may inhibit them from supporting proposals to deepen EU integration.

In external affairs the new members are likely to push for more EU attention to Russia, Central and Eastern Europe and the developing world. The newcomers share the goal of early membership for the Central and East European countries. Sweden and Finland both meet the UN aid targets of 0.7% of GDP for development assistance. All the new nations have an impressive record of contributions to UN peacekeeping.

Eastern Europe Waiting in the Wings

At the Essen European Council meeting in December 1994, the heads of state and government approved the Commission's plan for a White Paper setting out the requirements for the associated countries to integrate themselves in the single European market. They also gave their blessing to plans for an enhanced dialogue with the associated countries on a wide range of policy areas from security questions to the environment and transport. The Council ducked the tough problems, however. The Commission was to study how the costs of integrating the Central and East European countries would be met, and how to reform the CAP and structural funds.

The EU's magnetism, and the swiftness of the EFTA enlargement, has already fuelled speculation about the next wave. Both Poland and Hungary applied for full EU membership soon after their association agreements came into effect on 1 February 1995. The other associated countries (the Czech Republic, Slovakia, Romania and Bulgaria) can be expected to follow this pattern during the course of 1995. During 1994 free-trade agreements were signed with Estonia, Latvia and Lithuania, and in December negotiations started on association agreements. In March 1995, Italy finally dropped its objections and agreed that negotiations on an association agreement could start with Slovenia.

This means that ten more countries could join the EU in the first decade of the next century. In Copenhagen in June 1993 the European Council made the historic statement 'that the associated countries of Central and Eastern Europe that so desire shall become members of the European Union. Accession will take place as soon as an associated country is able to assume the obligations of membership by satisfying the economic and political conditions required'. These requirements included stability of political institutions, a functioning market economy, ability to cope with competitive pressures and acceptance of the EU's objectives of political, economic and monetary union. The EU would have to be able to absorb new members.

One of the first actions taken under the CFSP related to Eastern Europe. The Pact on Stability in Europe, proposed in 1993 by French Prime Minister Balladur, is an attempt to encourage the East European associates and the three Baltic states (but, controversially, not Albania) to reach among themselves a series of bilateral and multilateral agreements on respecting borders and protecting the rights of minorities, thus settling any potential security problems before they join the EU. Although the initial proposal of making EU membership conditional on signing such 'good-neighbour' agreements was dropped, the East European states have still been under the impression that participation in the Pact is mandatory for would-be EU members.

The inaugural conference of the Pact was convened in May 1994. A series of bilateral and round-table talks was then held and the Pact, a collection of the resulting agreements, was signed by the participants on 20 March 1995 in Paris. The Organisation for Security and Cooperation in Europe (OSCE – formerly the CSCE) is supposed to oversee the Pact's implementation. The thorniest issue was that of the large Hungarian minorities in Slovakia and Romania. Hopes that agreement would be reached were soon quashed. Hungary and Slovakia appeared to have made progress when agreement was supposedly reached on 19 March, but in speeches at the conference the next day it became clear that the two countries held different interpretations of what had been accomplished. Hungary and Romania only promised to continue with negotiations on an agreement, but prospects did not look good; Romania's prime minister did not use the word 'minority' once in his address to the conference.

The former Soviet republics are not included in the EU's enlargement plans. Instead, in 1994 and early 1995, the EU signed Partnership and Cooperation Agreements with Russia, Ukraine, Moldova, Kyrgyzstan and Kazakhstan, which provided for extensive ties between these countries and the EU. The EU has delayed signing a related trade agreement with Russia, as a result of the Russian involvement in Chechnya which awakened fears of a return to authoritarian rule in Moscow.

The Run-up to the Inter-Governmental Conference

Essentially the question for the IGC that is to be held in 1996 is how to organise a Union of perhaps over 20 members on the basis of democracy, transparency and efficiency. Both constitutional and institutional questions will have to be answered. Each new accession increases the burden of work and the diversity of issues to be handled. This suggests that the IGC will have to pay more attention to the application of the subsidiarity principle: what should be the balance between decisions taken and tasks attributed at the EU national and regional level? How can the Union ensure greater involvement in, and acceptance of, its activities by its citizens? Attention will especially need to focus on how to improve the preparation, taking and implementation of decisions in an enlarged EU which will inevitably involve more complex and diverse considerations.

The opening shots in this debate were fired by Karl Lamers, Foreign Affairs spokesman of the German CDU. He proposed a 'hard-core Europe' with Germany and France at its centre. UK Prime Minister Major responded in a speech in Leiden in September with calls for a more flexible '*à la carte* Europe'. French Prime Minister Balladur has spoken in favour of 'géométrie variable'. Whatever description is accepted there is common agreement on the need for substantial reforms affecting the Council, Commission and Parliament.

Perhaps the most sensitive issue is the question of changes to the voting system in the Council. The large states will certainly press strongly for greater attention to population size. Other contentious issues include the nature and composition of the six-monthly rotating presidency system and the troika (which consists of the past, present and future presidents plus the Commission); the manner in which the Council transacts business; and the EU's many official languages and consequent enormous interpretation and translation costs.

There is a wide consensus on the need to reduce the number of Commissioners, but there the consensus ends. If rotating constituencies cannot be agreed, the option of denationalising the Commission and drawing the Parliament into the nomination process may be explored. The fiasco surrounding Santer's appointment has certainly fuelled calls for a more open and democratic system to choose the Commission President.

The agenda for 1996 is lengthy, and the issues it will deal with are complex. Whether the deliberations can be concluded within a year is very much in doubt. The outcome will depend heavily on the political environment at the time and in particular on the occupants of the Elysée Palace and 10 Downing Street. No further enlargement of the EU will be possible until there is an agreement on institutional reform at the IGC. The 1996 rendezvous will thus have major implications for the future course of European politics.

Listening to Different Drummers

Three interrelated issues remained at the centre of European security in 1994: the proposed enlargement of NATO; the continuing dilemma over how to respond to events in Bosnia; and the search for a European security identity. Fundamental policy differences between Europe and America on these issues greatly strained transatlantic relations.

The proponents of NATO enlargement, apart from the countries in Eastern and Central Europe, included some voluble members of the Republican Party in the United States and Volker Rühe, the German Defence Minister. They argued that NATO enlargement would provide crucial security for the nascent democracies in Central Europe. Opponents argued that any NATO enlargement was bound to antagonise Russia, create new divisions in Europe, and lead to a weakening of the reforms in Moscow, without adding to the security of existing alliance members. To paper over these differences, the January 1994 NATO summit launched the PFP programme with the countries of the former Warsaw Pact. This was essentially a holding operation buying time until after the next round of presidential elections in the US and Russia. The sensitivity of NATO enlargement to Russia was underlined in President Boris Yeltsin's outburst at the Budapest CSCE summit in December and Foreign Minister Andrei

Kozyrev's refusal later that month to sign the proposed PFP programme for Russia. This was not enough to dissuade US President Bill Clinton from reiterating that no state would be allowed a veto over enlargement, which, he said, was 'not a question of if, but when'. The December North Atlantic Council meeting agreed to establish a study group to consider the criteria for enlargement.

As regards the former Yugoslavia, peace talks between the warring parties and with various outside bodies continued throughout the year with little progress. At its January 1994 summit, NATO declared itself willing to carry out air strikes in order to prevent the strangulation of Sarajevo, the UN 'safe areas' and other threatened parts of Bosnia-Herzegovina. In the last week of February, four Serbian aircraft were shot down near Banja Luka in the first aggressive military action by NATO since its formation in 1949.

The Balkan imbroglio took a further twist in February 1994 when Greece unilaterally closed its border with Macedonia and imposed trade sanctions on the republic as a result of the ongoing dispute over the name, flag and symbol proposed by Skopje. The Greek move was criticised by other EU member-states and the Commission, which took the Athens government, then holding the EU Presidency, to the European Court of Justice for breaching EU trade laws.

The Contact Group (the US, Russia, the UK, France and Germany) proved unable to hide their differences of approach towards the Balkans. US–European tensions over the arms embargo sharpened in the autumn and exposed deep divisions in the Group. In November, under pressure from a US Congress threatening to lift the arms embargo on Bosnia, President Clinton announced that the US would no longer participate in the international effort to enforce the arms embargo; US warships would not stop ships in the Adriatic suspected of carrying arms. The US decision immediately strained relations with the European allies. This, combined with America's apparent unwillingness to cooperate on a multilateral basis, could stimulate the creation of a European defence identity based on the WEU. The WEU, for example, immediately declared that the US decision on the arms embargo was deplorable.

A vigorous Anglo-French diplomatic offensive succeeded in convincing the US administration, but not Congress, of the merits of retaining the arms embargo. In December, the US agreed that if it became necessary it would provide troops to cover a withdrawal of UNPROFOR from Bosnia and Croatia, which helped to reduce tensions somewhat. But the transatlantic strains could prove difficult to repair. The new Republican majority in Congress, elected in November, fuelled concerns about possible US isolationism or unilateralism. The acrimonious visit to Europe by Senate leader Robert Dole soon after the elections only added to the concerns.

In the shadow of the battle between the US administration and Congress for control of foreign policy, the need for closer EU defence cooperation was clearer than ever. One sign that the European allies could reach a consensus on the possibilities and limitations of common defence was the positive reception of the 1 March 1995 UK paper spelling out its modest proposals for strengthening the WEU and defence cooperation in preparation for the EU's 1996 IGC.

Progress towards actually establishing a European Security and Defence Identity (ESDI) was hampered, however, in part by the problems emerging over the implementation of the Combined Joint Task Force (CJTF) framework. The CJTF idea had been launched at the January 1994 NATO summit. WEU forces would be able to use NATO assets (logistics and intelligence) in out-of-area operations. CJTFs could also include troops from non-NATO members. But the US insisted that all NATO members should approve each CJTF and that the US should approve the use of NATO assets. All CJTFs would have to pass through SHAPE (NATO's headquarters) and the Supreme Allied Commander. France objected vigorously to this: why should the US have a say in a CJTF if American forces are not involved? what if the US vetoed the use of NATO assets or the creation of a CJTF? The US has responded by reiterating that the US would most likely agree to a CJTF and the use of NATO assets. But the unilateral US decision to back out of a NATO policy to enforce the embargo against the former Yugoslavia destroyed the trust that lending such assets requires.

During the year, the WEU established a humanitarian task force, enlarged its planning capacity and sent policemen to provide for the security of Mostar in Bosnia. But the French-led efforts to secure an independent satellite capability for the WEU received mixed support. Another step toward enlargement was taken in May when the WEU approved the creation of the status of 'associate partner' for the six countries of Central and Eastern Europe having associate agreements with the EU, and for the three Baltic states. Associate partners would not benefit from the Article 5 guarantee of defence in case of an attack, nor could they vote on decisions with security implications, but they could participate in regular meetings of the WEU Council and take part in humanitarian and peacekeeping missions.

Looking Ahead

Although there was much vocal commitment to the ESDI, defence expenditure continued to fall in all European countries with the exception of France. The French government published a defence White Paper recommending policy continuity, but also holding out the prospect of closer ties to NATO and its major European partners. In October, France's defence minister attended a NATO defence ministers' meeting, breaking a boycott that had been in place since 1966. Germany also published its first White Paper on security policy since 1985, which focused on the reorganisation of its armed forces and cooperation with its European partners. Perhaps of more importance was the German Constitutional Court's historic judgement in July 1994 confirming that the government could send German troops abroad to participate in military missions as long as each venture was approved by the Bundestag. This important decision, welcomed by both the government and the opposition Socialist Party, clarified the German position, although it will be a long time before it is acted on. Both Chancellor Kohl and Foreign Minister Klaus Kinkel were quick to state that there would be no sudden rush to send German forces abroad, though a commitment to participate in a NATO-supported withdrawal from the former Yugoslavia was given.

The trend towards closer defence cooperation continued with Britain and France agreeing to establish a joint command to coordinate air force cooperation in humanitarian and peacekeeping operations. The Netherlands and Germany signed an accord for a joint army corps under NATO command, to be based in Munster. France, Spain and Italy pursued plans for a joint air-maritime brigade in the Mediterranean. The Belgians and Dutch agreed to establish a joint naval headquarters. But little progress was made on the thorny question of joint arms procurement, which would make a large, practical contribution to closer defence cooperation.

Defence will most certainly be on the agenda of the EU's 1996 Inter-Governmental Conference but, like the question of drastic institutional reform, it could prove so divisive that the status quo will be maintained. Those member-states most frustrated by the situation may then decide to leap towards much closer integration. The neutral states, plus Denmark, might object to moves to strengthen the WEU and to link it closer to the EU; other member-states, especially the UK, could block any moves to entrust the EU with greater responsibilities in the defence field.

Closer defence cooperation will be hotly debated as the IGC approaches, and a sense that the US will take a more relaxed attitude towards such integration – even encourage it – may spur some countries on. But if there is one area where *à la carte* choice is most likely, it is in the defence field. Here the appearance of a coalition of the 'ready and willing', or their absence, is going to be more important in the near term than the development of alliance-wide structures in the WEU.

Time for Peace in Northern Ireland?

On 23 August 1170, Richard, Earl of Striguil and Pembroke, known as Strongbow, landed at Waterford to do battle with Rory O'Connor, high-king of Leinster – and one of the longest running struggles in modern history was under way. On 31 August 1994, some 824 years to the month after Strongbow's landfall, and 25 years to the month since the First Battalion of the Prince of Wales' Own Regiment was deployed in the Catholic Bogside district of Belfast, a cease-fire was ordered by the Provisional Irish Republican Army (IRA) in their campaign to end British rule in Ulster. When Loyalist paramilitary groups, recently even more active in murder than the IRA, followed suit with their own truce, the threat of military violence in the province was lifted for the first time in a generation. British soldiers no longer took to the streets in full battle dress; patrols were cut back and planning began to reduce the overall military presence; negotiations were scheduled between veteran IRA insurgents and the authorities which had once silenced, exiled or even jailed them. Some began to wonder if they were witnessing the beginning of the end of the long history of the troubles in the island.

Inevitably, this historic turn of events has raised two questions: what brought about this outbreak of peace? And how can it mature into institutions for the peaceful management of Northern Ireland's divided loyalties?

From Stalemate to Negotiations

The IRA cease-fire was greeted by many as a turning-point in Anglo-Irish affairs. It had always been up to the IRA to call the war off, or to keep up the pressure. The IRA's Army Command, a committee of seven senior officers directing the insurgency, and its political front organisation, Sinn Fein ('Ourselves Alone), had come a long way since the so-called 'Troubles' flared up again at the end of the 1960s.

At that time, the IRA was a spent force, its Dublin-based leadership under Cathal Goulding more interested in building bridges to working-class Northern Protestants than with terrorising their middle-class leaders in Stormont – the 'Protestant Parliament in a Protestant state', as Loyalists called it, which ruled the province within the United Kingdom. It has been claimed that there were fewer than 60 IRA cadre in Belfast in the late 1960s, half of them inactive. A frantic search by the IRA for weapons in Belfast at the time revealed a motley assortment of only 75 guns, mostly sporting pieces, only about a third of them serviceable.

By 1992, however, the IRA had arisen from the ashes like the phoenix, the emblem of its predecessor, the Fenian Brotherhood (motto: 'revolution sooner or never'). Re-formed in Northern Ireland as the Provisional IRA under Sean MacStiofain (born John Stephenson in England), it had evolved into a clandestine force with a few active service units operating on the British mainland. It had also learned the art of public relations, winning sympathisers in the United States and support from such patrons of troublemaking as Libya and Iran. Sinn Fein even won elections to local and national office.

By 1992, moreover, its tiny armoury of handguns and shotguns had reportedly grown to include 650 AK-47, 20 DSHK Russian heavy calibre armour-piercing machine guns, one SAM 7 surface-to-air missile (nine others had been recovered by the authorities), 40 RPG rocket launchers, 600 detonators and 3 tons of Semtex explosives – about half of a shipment from Libya six years before. Various estimates at the end of 1994 put the number of IRA activists between 200 and 500, to which, perhaps, should be added many of the 700 claimed by Sinn Fein to be Republicans serving sentences in prisons in Great Britain and throughout the world.

The cost of the IRA insurgency to the governments and people of both Ulster and Great Britain has mounted up over 25 years: overtime for the Royal Ulster Constabulary (RUC) alone costs £82m a year, with an annual estimated cost to the British taxpayer of about £8bn, including subsidies to shore up the local economy and maintain approximately 18,000 troops on active duty. The one-ton bomb detonated on 24 April 1993 at Bishopsgate, London, wreaked extensive damage, estimated at upwards of £1bn. But it is, of course, the gruesome record of casualties on all sides which has kept Northern Ireland a focus of concern, not only in London and Dublin, but across Europe and the English-speaking world.

From 1969 to 1994, approximately 3,150 people died as a result of violence in Northern Ireland. The victims ranged from the former Shadow Northern Ireland Secretary, Airey Neave, and the Earl Mountbatten, Prince Philip's uncle and a confidant of the royal household, to nine-year old Patrick Rooney, killed in his bed by a stray bullet from a .30 calibre machine gun used by the police to fire on residential blocks suspected of sheltering snipers. The toll nearly included former Prime Minister Margaret Thatcher and most of her cabinet. Hundreds of innocent by-standers, Catholic and Protestant, died as IRA bombs detonated accidentally, gunman killed the wrong people by mistake, or bombers targeted public places, claiming them to be military targets. For their part, Protestant paramilitaries at times did not even bother to deny the purely sectarian nature of their killing of 'taigs' (Catholics) with no IRA connections, presumably to inspire fear among the Republican population.

Some of England's proudest legal traditions fell victim to the deterioration of order in this other island: London was forced to impose internment (later called 'preventive detention') at one point, institute trial without jury and limit free speech and movement by individuals. It even became implicated in, or at least suspected of, police practices which pushed its widely admired criminal justice system to the brink of disrepute. British policy could take comfort from the fact that the great majority of Catholic, and presumably Republican, sentiment in Ulster backed the Social Democratic and Labour Party (SDLP), not the IRA. Unable to win a military victory, IRA strategy has been to 'sicken the Brits' by urban warfare intended to wear down Westminster's resistance until it is ready to sacrifice the desires of Northern Irish Protestants for the sake of peace and quiet. Like insurgents everywhere, the IRA has been nurtured on, and often the prisoner of, the lessons and mythologies drawn from its struggle: sacrifice and endurance are the secrets of victory; compromise is defeat; popular support flows from military success; only physical force can succeed.

From Negotiations to Cease-Fire

If little else, 25 years of stalemate showed that each side could persevere in its strategy. The IRA was unable to force London to betray its commitment to Ulster's Protestant majority, while Great Britain could not prevent a small group of dedicated men and women from perpetrating atrocities.

Key landmarks on the road towards peace were the joint policy statements by Dublin and London in the Sunningdale Agreement of 1974 and the Anglo-Irish (Hillsborough) Agreement of 1985, both asserting that constitutional change in Northern Ireland could not come about without the consent of the majority, but both providing for an increased role by the Republic in the province's affairs. The first was overturned by a general strike of hardline Unionists, alarmed that plans for greater North–South consultation amounted to unification by stealth. The second never got off the ground. Both were ignored by the IRA as it pinned its strategy to 'one big thing' – a non-negotiable demand for British withdrawal.

An end to this impasse was slow in coming. By the end of 1993, however, things seemed to be stirring once more. According to the British government, communication began with an oral message from the IRA on 22 February 1993 stating that 'the conflict is over', and asking London's advice on means to bring it to a close. (The IRA denies it made such a statesman-like overture, which might, after all, be interpreted as suing for peace.) Revelation of contact between the IRA and London drew predictable outrage from many Unionists, who had argued consistently that Republican violence was a criminal matter and there was nothing to talk to the IRA about. Yet almost as significant as these contacts was how restrained general public reaction was to them.

'The message is clear,' observed *The Guardian*, 'People may not like it, but they will swallow it.' A readiness to talk, moreover, may have been encouraged by the disintegrating situation on the streets of Northern

Ireland. Beginning in late October 1993, a new wave of random violence resulted in 20 deaths in eight days and, after a brief pause, produced four more murders within 24 hours. The statistics themselves were beginning to convey a different story than British opinion had been exposed to in the past. In 1993 the Troubles resulted in a total of 84 deaths, 50 of them claimed by 'loyalist paramilitary forces', and 34 attributable to 'Republican terrorists'. For the third year running, Loyalist killings outnumbered those of the IRA. But for the first time since 1969, the British Army and the security forces had killed no one.

There were, moreover, too many other developments under way to balk at constructive secret diplomacy by Westminster. Between April and September 1993, Gerry Adams and Derry SDLP MP John Hume (who had been in contact as far back as 1988) explored the terms on which violence could end and Sinn Fein could be admitted to the political process. The Irish Foreign Minister, Dick Spring, announced his own six-point plan while Northern Ireland Secretary Sir Patrick Mayhew continued with his 'three-strand' programme of consultation with the constitutional parties and cross-border political institutions.

From Downing Street to Stormont

In the closing weeks of 1993, what might be called the endgame of Anglo-Irish diplomacy came in the form of the Downing Street Declaration, issued by the Irish Taoiseach, Albert Reynolds, and the British Prime Minister, John Major. Termed a masterpiece of ambiguity by one Unionist, it set the agenda for progress during 1994 by drawing together the strands of negotiations which had been carried on for some time through secret and not-so-secret contacts.

The British government declared it had no 'selfish strategic or economic interest in Northern Ireland', but would protect the democratic right of the majority there to say where its loyalty lay. This was an important signal to the nationalists that British policy was not a matter of imperial claim (as most IRA sympathisers are brought up to believe) but of constitutional obligation; self-determination in the province, not union with Britain, was its guiding principle. For his part, the Taoiseach acknowledged that despite assurances on their constitutional status, Ulster Unionists feared for the integrity of their traditions, and that such fears must be dealt with if progress was to be made. Thus he undertook to work for changes in the Irish constitution, which claims sovereignty over the North.

Like Sunningdale and Hillsborough before it, the Declaration anticipated the progressive growth of all-Ireland institutions; it also specifically urged that 'old fears and animosities' be replaced by a readiness to negotiate 'which would compromise no position or principle nor prejudice the future of either community'. It held out the prospect that Republicans, even veteran insurgents, could keep their long-range, nationalist goals as long as they obeyed the law. In the meantime they were being asked to cooperate with a democratically elected legal authority, not to be eternally loyal to it.

The ball was now in the IRA court: it had to choose between the futile mystique of armed struggle, and the uncertain but realistic rewards of political struggle. Its repeated requests for 'clarification' of the Downing Street Declaration were seen as stalling by many, and Unionists were outraged that the British government should even consider replying. But reply it did, at the end of May, thus closing one more avenue of retreat by the IRA from confronting a decision to end violence. Reassessing policy, however, does not come easily to the cadre of such a movement. To the despair of many, the Sinn Fein *ard fheis* (annual conference) at the end of July rejected the Downing Street Declaration and Gerry Adams refused to recommend a cease-fire: the stalemate seemed destined to continue.

Not two weeks later, however, Adams was again hinting at a cease-fire, and by the end of August, even the notoriously hardline MacStiofain announced that it was time to end the policy of violence. An IRA cease-fire (that is, a cessation of military operations) was finally called for at mid-night on 31 August, without time limit, but noting that the Downing Street Declaration was 'not a solution' and that the IRA had not abandoned its Republican commitment.

Now it was the turn of the British to appear to be stalling as Prime Minister Major demanded that the cease-fire be characterised as 'permanent' – even though the IRA leadership may not have been empowered to do so, nor able to bind their successors any more than the British parliament could. Eventually, it became possible, indeed inevitable, to regard the cease-fire as a 'working assumption'. By early December 1994, Sinn Fein had named its representatives to attend Stormont meetings to begin in the New Year.

On 22 February 1995, a little more than a year after the Downing Street Declaration and with six months' peace in the province, the British and Irish governments put in place a Framework Document to provide an Anglo-Irish agenda for all-party negotiations on the future. The document anticipated the return of a large measure of devolved self-government to the province, but under a new form of administration and with new safeguards for the minority.

A legislative assembly, elected by proportional representation, would supervise the executive function of the state through committees, whose chairmen would act as heads of government departments. A novel element was the idea of a panel, possibly of three members, elected by proportional representation, with a wide remit including the supervision of legislation, approval of certain appointments, fiscal oversight and – most critically – nomination of the committee chairmen who would constitute the province's executive. Since these last would be chosen in a manner 'broadly reflecting proportional party strengths in the Assembly', the panel could be intended to apply the long-mooted ideal of power-sharing between the communities. The sensitive area of policing is not ignored, with the suggestion that consideration be given to promoting the degree to which 'the community at large . . . can identify with and give full support to the police services'.

The Framework Document also elaborated on the Downing Street Declaration's generalised goal of greater North–South cooperation. Confirming yet again that constitutional change in the North (i.e., change of sovereignty) could not take place against the will of the majority, it offered a number of phrases which no doubt had many Unionists reaching for their correcting pens: a call for 'structures . . . to exercise shared administrative and legislative control over [agreed] matters'; North–South institutions 'to carry out . . . delegated executive, harmonising and consultative functions'; and a 'new approach to traditional constitutional doctrines on both sides', among other things.

Unionist reaction to the Framework Document was predictable and immediate – from expressions of grave disappointment to incensed rejection. Despite the vigour of the response, however, there were no immediate threats to settle the matter in the streets, as had happened in 1974. Unionism, in its purest dispensation, stands for the *status quo* and must logically reject any change as an erosion of the existing and preferred constitutional status. In the past, it has sought to shelter such *immobilisme* behind a democratic totalitarianism which provoked its enemies and ultimately alienated its friends.

The apparent IRA turn from violence has now robbed Unionism of an enemy long shared with Great Britain and which had provided a common cause between London and Belfast. In the wake of the events of 1994, it stands today at a crossroads, reluctant to try new methods of government based on cooperation rather than political monopoly, but unable to control its destiny on its own terms.

What is striking, and must be sobering to Unionists, is the consistency of the British policy from the beginning to the end of the present troubles. Across 25 years of conflict both Sunningdale and Downing Street asserted the same principles of democratic choice of constitutional status, power-sharing, and development of closer North–South relationships. For extremists at both ends of the Republican and Unionist spectrum, it must increasingly seem that the Framework Document is the only option. Now that the IRA has opted into the political process, it will become extremely difficult for the Unionists to opt out.

When the Shooting Stops

Some of the explanation for the IRA decision must be fatigue – weariness with a battle that has gone on for a generation without any perceptible territory or trophy to show for the sacrifices. The leaders of the IRA or Sinn Fein now see their own children embarking on what might be another generation of violence, with all its new risks (the son of one of the IRA negotiating team, Sean McManus, was killed in an IRA ambush). The life of an IRA activist is certainly mean and brutish, and may often be short. It involves poverty (a pittance of subsistence, and none if the man is drawing the dole) and violence which can turn idealists into murderers and young men into middle-aged convicts who have passed an adolescence and youth in maximum-security prisons – if they are lucky. Some of its operations have been known to sicken its own troops as much as the British.

Time, moreover, was not on the side of the IRA. While it certainly commands greater resources now than when it was recreated in 1969, its numbers are thought to have fallen in recent years, and funding is increasingly a problem. Nor is it capable, in the circumstances, of adequate levels of training for clandestine efficiency, as its numerous mistakes and bungled operations demonstrate. Above all, it is badly outnumbered. If, as some claim, successful insurgency requires a ratio of 1 activist to 30 soldiers or police, the IRA force, with odds at 300 to 1 against, is in a hopeless position. In addition to regular British soldiers, there are potentially 20,000 armed Protestants, a force of 12,000 RUC, and the miscellaneous Protestant paramilitary forces already active against them. The IRA can continue to fight, but it cannot win.

Nor is the IRA/Sinn Fein in a position to claim the loyalty or even much support from the population whose defender and nationalist conscience it claims to be. In the last democratic elections it polled less than 2% in the Republic and about 10% in all of the North, representing only one-third of the Catholic vote, which is preponderantly cast for the SDLP. The IRA may tap some nostalgic sympathy from the Catholic community, but if anything is clear in Northern Ireland, it is that the IRA campaign of violence has alienated it from its own community and deprived it of the grass-roots support which Michael Collins, the hero of Irish independence, memorably demonstrated is the vital terrain of any guerrilla campaign.

A sense that the time for change had arrived in Ireland, as in South Africa, Eastern Europe and the Soviet Union, may also have had some influence. The end of the Cold War marks a turning-point in history, one which broadcasts its lessons of inevitable change and irresistible evolution even to those far removed from its central narrative. The peaceful revolution in South Africa and the long delayed Arab–Israeli agreements demonstrated that what seems impossible can happen as a result of human choice. The present Troubles in Ireland began as the social restlessness of the 1960s overturned the orderliness of the early post-war world. Their end may complement the other great shifts in world affairs that are shaping the closing decade of this century.

The new formulation of the Irish question today is whether the pace and direction of current developments can be sustained and built upon for a lasting peace, or whether they add up, in the end, to little more than another pause between campaigns. That the process of talking, the fundamental transaction of all civil society, has finally begun offers the first basis for optimism. It is often noted how situations irretrievably degenerate into conflict and how events propel states into confrontation almost against their will. But it is also true that peace carries its own momentum, gaining in force from an accumulation of even small victories: breaking off talks can often be far more difficult than starting them. For the IRA to resume military operations – or for any party to be seen as causing talks to break down – would take a great deal of explanation to all concerned: prisoners hoping for release, voters, active cadres, journalists, foreign governments and even long-standing supporters.

In the forefront of those asking difficult questions would be the investors. With the stalemate broken by the IRA cease-fire, a heavy bombardment of cash for Northern Ireland was laid down to support the peace offensive. EU President Jacques Delors proposed an additional £233m grant on top of the approximately £1bn presently provided; major investments were under consideration by DuPont, Ford and British Telecom, among others. Renewed violence would scatter these investments, their promise of jobs, to the four winds; even hardened activists shrink from causing such damage to the province's interests.

There are, moreover, many signs that the situation today is not that of 1974, when Sunningdale was brought down by hardline Protestant forces. The Reverend Ian Paisley aside, Unionist opinion has mellowed and is more ready to distinguish between what it will not surrender and what it is prepared to accept.

One of the most striking signs of hope was the declaration of a cease-fire by the Combined Loyalist Military Command on 13 October 1994. Cleared in advance with Protestant Loyalist convicts serving sentences in the Maze Prison, it contained what seemed to be a heartfelt, and uncharacteristic, expression of regret for the hurt which had been caused over the years. Protestant paramilitaries have been quoted as saying that they regard the Downing Street Declaration as satisfying their political requirements, and at a paramilitary rally at the end of October one speaker urged 'we must ensure that the mistakes and injustices carried out in the past in our name never happen again'.

Newspaper reporting from the province has suggested that the attitudes expressed in the pubs in even Loyalist strongholds suggest more open-mindedness towards the peace process than might be assumed from the rhetoric of Loyalist politicians. There seems some popular acceptance that change is coming and it is better to take part in shaping it than to accept what others make of it.

It is unlikely that Unionist control of Northern Ireland would attempt to reimpose the political and social techniques of discrimination which once marked Protestant majority rule in Ulster. Considerable strides have been made to introduce fair employment practices and break down the sectarian exclusiveness which once dominated public life. In October, the Northern Ireland Police Authority for the first time demanded assurances from its Chief Constable of fair employment practices. There has even been some discussion of dispensing with the oath of allegiance to the Crown for senior positions; it is already unnecessary for many minor posts.

For its part, the IRA may finally be displaying a readiness to confront and take seriously the concerns of the Unionist community in the North – concerns it had often brushed aside as irrelevant to the only real issue, which in its eyes was its struggle with the British government. There must be a dawning realisation that no British government is interested in re-maining in Ulster one minute longer than protecting the interests of the majority commits it to. This in turn suggests that the task of coming to terms with their Unionist neighbours presents a more politically signifi-cant challenge for Republicans than sniping at policemen and bombing pubs in Surrey.

To be sure, the future will not be an easy one for any of the parties, and difficult issues remain to be settled before a new page in Irish history can be turned. Suspicious Unionists will scrutinise attempts to develop institu-tions for all-Ireland co-operation, ever careful to detect the slightest depar-ture from the claim of absolute British sovereignty in the North. IRA veterans may discover how frustrating politics can be particularly when compared with the cathartic, if futile, resort to violence.

There are, moreover, a number of issues for which cooperation and goodwill are essential if progress is to be made. They include the problem of the IRA arsenal – how it can be deactivated to the satisfaction of Great Britain and when can it be abandoned with security by the IRA. The treatment of prisoners – IRA and Loyalist – will certainly pit strong emo-tional commitments against unyielding legal principles: some arguing that amnesty cannot be granted to murderers; others arguing that prisoners of war (as the convicts see themselves) should not be held after the fighting is over.

The politics of Northern Ireland, moreover, is not an Irish matter exclusively, any more than it was in the days of Disraeli and Gladstone. Prime Minister Major's majority is thin, and his party is bedevilled over

Europe. It is true that Unionists can hardly look for support to the Labour opposition, but they have already indicated that they can no longer be counted on to back the government as an ally. Worse still, pro-Unionist sentiment among Conservative backbenchers might still surface to add new threats to Major's party leadership, if not his parliamentary majority, and thus challenge his hopes of a historic, personal triumph as the statesman who finally brought to a close one of the longest running struggles in the Western world.

Economic factors are also involved in any permanent solution: unemployment and de-industrialisation in the North, if uncorrected, will continue to exacerbate urban resentments. Finally, many deep and unhealed wounds remain from decades of killing; there are many vested interests in preserving the power to intimidate a neighbourhood and exploit violence; the RUC is still 93% Protestant and Catholics still top the unemployment statistics. Both communities cling to powerful emotional attachments to their histories, and there is little disposition to forget the heroics, and injuries, of the past.

Above all, of course, is the abiding difference in the sense of national identity. There is more reason now to hope that while such conflicting allegiances explode in violence elsewhere in Europe, in Ireland, where they have so long dominated the agenda, it has finally been recognised that divided loyalty is simply another one of the many differences people can learn to live with – if they are ready to substitute the techniques of democracy for those of the parade ground. The first steps in this direction were taken in early 1995. A few more strides in the same direction should produce a momentum that will assure the leaders of all the parties concerned the reward promised to peacemakers.

There is still much to ask. London may need to revive the constructive attitude that earlier British governments had demonstrated towards their former enemies. In the IRA it will require courage and leadership to convince the hardliners that politics is a human and thus imperfect arrangement, that compromise is not betrayal, and that violence is not necessarily honourable. Unionists will have to accept that the majority has duties as well as rights, and these rights are not themselves without limit. What is encouraging is that, at long last, it is not foolish to ask.

The Middle East

The Middle East peace process teetered between hope and despair during 1994. Israel signed a peace treaty with Jordan ending their 46-year state of war, but seemed unable to push its relations with the Palestinians out of deadlock. The continued terrorist attacks by extremists played havoc with Israel's hope that security would be attained by completing the Oslo agreement, leaving the Rabin government doubtful about making the further concessions necessary to move beyond the halfway point that has been reached. Despite consistent efforts by the United States, President Assad of Syria still felt that he could gain more by prevaricating on peace than by accepting the options offered by Israel and the US. With elections looming in both Israel and the United States in 1996, there was a growing sense that unless a way can be found to advance the peace efforts before the end of 1995, the best opportunity for peace that has appeared since 1949 will be lost.

Radical Islamists were not only active against the Israelis. In Algeria, faced by an intractable military government that had refused to honour their leading position in a democratic election, frustrated fundamentalists turned their guns against police, judges, journalists and foreigners in a campaign to render the country ungovernable. They have largely succeeded in this aim, but this success has not been reflected in political compromises by the country's leaders. The leaders of Egypt and Saudi Arabia, also facing an awakening rebellion, have fared better in their efforts to dampen the appeal of the intransigents. Saudi Arabia has not only faced domestic unrest, but was also concerned about troubles on its southern flank, in Yemen. And the Gulf's pariah, Saddam Hussein, continues to try to split the West in his efforts to recover a more commanding position for Iraq.

Although no one believed that the countries of the Middle East would soon thrive in an atmosphere of tranquillity, there had been an expectation that they were on their way to a greater degree of normality than had been seen for generations. The progress that has been made, however, has been so slow and hesitant that it has not provided a sufficient bulwark to prevent a reversion to mutual suspicions. It is as difficult to maintain an attitude of cautious optimism now as it was easy to do a year ago.

Ups and Downs in the Search for Peace

On 26 October 1994 Jordan and Israel formally ended a state of war which had lasted for 46 years. The signing of a peace treaty in the barren desert at Wadi Araba by King Hussein and Prime Minister Yitzakh Rabin, with US President Bill Clinton beaming in the background, was the most important development during the year in relations between Israel and the Arab states. Jordan thus became the second of the four Arab confrontation countries bordering Israel (Egypt, Jordan, Syria and Lebanon) to move out of the circle of official war. That Jordan was willing to do so reflects the extent to which the agreements between Israel and the Palestine Liberation Organisation (PLO) reached during the preceding year have diminished the long and bloody Arab–Israeli conflict. It also reflects the change in Arab foreign policy that now allows each state to act in its own national interests with greater independence from the rhetoric of 'Arab solidarity'.

But the Arab–Israeli conflict is not over. Serious differences remain among the major parties, while a deadly terrorist onslaught, instigated mainly by radical Islamic groups opposed to the peace, continues in Israel and against Jewish targets outside the Middle East. The prospects for a strong and prosperous Palestinian entity under Yasser Arafat are dismal. Without significant advances towards the final stages of the Oslo Principles, Israel may decide to separate itself physically from the Gaza Strip and the remnants of a Palestinian entity on the West Bank by constructing fences or barriers. Despite all this, both sides remain committed to negotiations, and the chances of achieving a reasonable peace are still real and reachable.

Wading in to Untested Waters

King Hussein's decision to sign a peace treaty at this stage took many by surprise. It had generally been assumed that he would not risk a move that Syria would see as another attempt to isolate and weaken it. It was thought that the King would prefer instead to reach a tacit understanding and agreement with Israel, leaving the signing of a peace treaty for a much later stage.

King Hussein appears to have been motivated by a concern that Jordan might fall behind Israeli–Syrian negotiations as well as by concern that Israel could make rapid progress along the Palestinian track, leaving Jordan with little or no influence over the West Bank. Most of the fruits of peace would thus fall to the Palestinians. Since Jordan's economy and its relations with Saudi Arabia had suffered following the 1991 Gulf War, this possibility was not an attractive prospect for Amman. That Jordan would also improve its relations with the US by taking this final step towards peace, and thus gain help in writing off its debts, added considerable weight to the balance of Jordanian calculations.

The peace treaty promised strategic advantages to both Israel and Jordan. Since Israel would henceforth have a vital interest in preserving Jordan's stability, Jordan would be able to relieve the pressure from its more powerful neighbour on its western flank. It would have a secure and delineated western border. Jordan has no longer to fear Israeli measures to force Palestinians out of the Occupied Territories into its land, danger-ously upsetting the demographic balance. Clause 6 in Article 2 of the peace treaty explicitly forbids moving a civilian population against its will and in a manner that puts at risk either party to the agreement. The treaty has thus dealt a blow to the concept nurtured by Israeli right-wing circles that 'Jordan is Palestine'. To drive the point home King Hussein and Crown Prince Hassan also met with the head of Israel's right-wing Likud Party, Benjamin Netanyahu.

In the wake of the agreement, Jordan occupies a more important position in the Israeli–Palestinian–Jorda-nian triangle. It can now feel even more sure that Israel will make every effort to protect it from harm. Israel's undertaking to do all it can in negotia-tions with the Palestinians to maintain Jordan's special position over Islamic shrines in East Jerusalem was an early demonstration of this new-found ad-vantage. The agreement with Israel provides Jordan with more water from the Jordan River; and will enable Jor-dan to reclaim territory in the Wadi Araba region of the Negev and in the north of the Jordan Valley which it lost after the 1967 war. The agreement has also produced American aid for Jor-dan.

While peace with Israel has clearly enhanced Jordan's security, it has also led to important changes for Israel. It had never been the size of Jordan's army that most disturbed Israel, but its geography. Jordan's army formed the central link in any eastern front. Its proximity to vital Israeli objectives (airfields, industrial and population centres) created consider-able anxiety in Israel whenever Jordan considered a change in military deployment. Israel always feared that Jordanian territory would be used for the deployment of, or as a passage for, the armies of an Arab military coalition. From its earliest days Israel has regarded any hint that either was under way as a *casus belli*. The agreement has dealt with the problem to Israel's satisfaction since both parties have undertaken not to host other forces as a secondary threat. Indirectly, the agreement has also relieved Jordan's dependence on Syria and prevents Amman from slipping into the Syrian orbit, as it slipped into the Iraqi orbit in the late 1980s. If Jordanian relations with Israel are consolidated they could mature into a strategic cooperation very different from the cold Egyptian–Israel peace.

Territorial Arrangements

The agreement contains provisions for returning territory to Jordan. But Israel scored well in not being compelled to dismantle the settlements near the border whose economy depends on these lands for water cultivation. Had the Rabin government been forced to dismantle settlements it would have met with strong public opposition. Israel and Jordan solved this difficult problem by including arrangements for Israel to lease two small pieces of territory (approximately 6km^2) within the clauses of the agreement covering the exchange of territories, encompassing a much larger area.

One piece of the leased territory had been bought by the Israeli electric company during the British mandate. All the territory has now been transferred to Jordanian sovereignty and Jordanian law will prevail. Despite this, much of the criticism of the agreement that has come from the Arab world, particularly from Syria but also from Egypt, has focused on the leasing clause for fear of the precedent it might set.

Good as it is, the agreement between Jordan and Israel has hardly resolved all their problems and resolution of some remains doubtful. One major issue is the continuing burden of refugees on Jordan. This problem has shifted to the quadrilateral talks which include Egypt and the PLO and which convened its first session in early March 1995. Some relief may be found for Jordan by facilitating the return to the West Bank of Palestinians uprooted by the 1967 war. Because of its interest in maintaining stability in Jordan and in preserving the Hashemite regime, Israel can be expected to do what it can to avert irredentist pressures on Jordan. It will do everything possible to strengthen its neighbour economically and will look with understanding on Jordan's defensive needs. Israel recognises that Jordan's army is an essential centre of gravity for the regime.

Even a successful peace treaty does not exonerate Jordan and Israel from tackling the serious question of how to integrate the Palestinians in a comprehensive peace settlement. It is in the interest of both countries to guarantee that the future Palestinian entity does not send destructive waves westwards to Israel or eastwards to Jordan. The Palestinians were critical of the peace treaty, although they are likely to gain from it. If the dangers to Israel from the east are genuinely minimised and Jordan no longer presents the threat of providing deployment space or transit facilities for armies seeking to attack it, Israel would no longer need a massive military deployment on the West Bank. Were Syria to join the settlement, Israel's needs for military deployment on the West Bank would be reduced even further.

The Lebanese Stumbling Block

The chances of a settlement between Israel and Syria must be viewed in light of events in southern Lebanon where a mini-war is in progress. The war pits Israel and its proxy, the SLA (a Lebanese militia under Israeli patronage), against the Shi'i militia commonly known as *Hizbollah*. The longer this war goes on the more momentum it accumulates and the more it threatens the negotiations. It draws the negotiators' attention away from the central issues to subordinate ones at the risk of losing control and causing a large-scale flare-up.

Israel is convinced that Syria, which for all practical purposes controls Lebanon through the massive forces it keeps there, wants the mini-war. It does nothing to initiate the clashes, but it does nothing to stop them. It also allows equipment and weapons to flow from Iran to *Hizbollah* in Lebanon via Damascus. Syria may regard the war as a useful means of exerting pressure on Israel. At the same time, Damascus played a reassuring role when Israel responded to the shelling of settlements in Galilee by launching a large-scale military campaign into Lebanon in early 1994. In the course of this drive some 200,000 or more inhabitants were expelled from their homes in southern Lebanon. They swarmed towards Beirut, creating strong pressure on the government. With US mediation and Syrian consent, an understanding was finally reached that in future Israel and *Hizbollah* would avoid striking civilian population centres, assuming that they are not the bases from which attacks are mounted.

This agreement has not stopped the fighting, which continues to cost lives on both sides. In the course of the fighting Israel has also struck *Hizbollah* bases in the heart of Lebanon. In the eye-for-an-eye, tooth-for-a-tooth structure of the Arab–Israeli conflict most observers are certain that, in retaliation for its losses in Lebanon, *Hizbollah* was responsible in July 1994 for the explosions at the Israeli Embassy in London and at the Jewish community centre in Buenos Aires which claimed over a hundred lives.

Israel has consistently declared that it has no territorial claims on Lebanon and that, provided certain conditions are met, it would withdraw from south Lebanon. The Israeli Prime Minister posited four conditions: the deployment of the Lebanese Army along the northern border of the security zone inside Lebanese territory; the disarmament of *Hizbollah*; negotiations after six months with no attacks on Israel, followed by an

Israeli retreat to the international border; merging the SLA, along with the many civilians, teachers, doctors and others who have been collaborating with it, with the Lebanese Army, as other militias have been.

In some ways this is a proposal for an interim settlement between Israel and Syria, with the focus on the situation inside Lebanon. Damascus has understandably been actively opposed, mainly out of concern that what Israel really has in mind is a separate peace with yet another Arab country. Such a move would isolate Damascus still further, given the peace treaty between Israel and Jordan, and the settlements between Israel and the PLO.

On Dead Centre

Direct and indirect negotiations between Israel and Syria last year yielded no practical results. Despite their efforts, the US mediators achieved no breakthrough between the parties. Twice President Clinton met President Assad, first in Geneva in January 1994 and then in Damascus after Clinton had participated in the peace signing ceremonies between Israel and Jordan, with the idea of moving the peace process forward. At a press conference in Geneva Assad declared that Syria regarded peace as a strategic objective. But he did not publicly denounce terrorism which is of primary concern to Israel.

Despite the positive nature of the meeting between Clinton and Assad, Washington has not struck Syria off the list of countries supporting terrorism. This position hardened when Islamic *Jihad* leader Fathi Shikaki announced from his home in Damascus that his organisation was behind the Beit Lid bomb attack on 22 January 1995, one of the worst terrorist acts perpetrated in Israel. In contrast, the European Union has agreed to lift the arms embargo that had been imposed on Syria when it was accused of involvement in a terrorist attempt against an El Al passenger aircraft about to take off from London.

Although there was little progress in the negotiations, the Israeli–Syrian ambassadorial-level talks continued under US patronage in Washington. For the first time the parties addressed security issues when Assad sent Chief-of-Staff Hikmat Shihabi to Washington in late December to meet Israeli Chief-of-Staff Ehud Barak and his team. The meeting was not really a negotiating session: each side simply presented its own position regarding the security arrangements that would have to accompany any agreement between the two countries. President Clinton invited both delegations to a meeting at which he stressed his country's interest in the success of the talks. At the end of the Washington talks, an Israeli representative told the Syrian Chief-of-Staff that he believed 'there was something to talk about and someone to talk to'.

Despite his approval of the meeting between the two chiefs-of-staff, Assad has consistently refused to meet Israeli Prime Minister Rabin. Rabin argues that no breakthrough is possible unless the two leaders hold direct discussions, while Assad counters that summit diplomacy is dangerous and that there could be serious repercussions if such a meeting were to end in failure. In Israel this is understood to mean that Assad is ready to meet Rabin provided Rabin relinquishes his negotiating positions beforehand. The chances of a preliminary meeting between the two leaders are very slender.

Israeli and Syrian positions diverge on almost every point. Israel insists that security requires it to remain in at least part of the Golan Heights; Syria insists on a total withdrawal. Israel particularly objects to withdrawal from the Hamma triangle and the Samakh triangle between the Yarmonk, Jordan and the Sea of Galilee, which were both seized by Syria during the 1948 war, but it is also thinking of making other border amendments. There is no agreement on the duration of withdrawal, although it is understood that it is to be phased. There are differences over the nature of normalisation (diplomatic relations, open borders, trade and tourism). There are differences over security arrangements, starting with the reduction of the military forces and ending with the size of the demilitarised zones, as well as over the question of the symmetry of Israeli and Syrian redeployments.

The stationing of an American force on the Golan Heights when agreement is finally reached, has become a controversial issue in the United States. With support from American friends, Israel right-wing activists opposed to an agreement involving territorial concessions have been lobbying the US Congress against deploying any forces. Like its predecessor, the Clinton administration maintains that if Israel and Syria ask the US to deploy forces to the Golan Heights to ensure that the agreement is put into practice, Washington would agree and ask for Congressional approval.

While these differences distract from any early agreement, there are other concerns for each party which push in the other direction. While President Assad may believe that he has captured the pan-Arab high ground, he must also be uneasy about the agreement that Jordan, and the PLO before it, hastened to sign with Israel. Nor can he be happy with the opening of an Israeli Office of Interests in Morocco and Tunisia, or the close ties which are developing with Oman and Qatar.

These moves to bring Israel and some of the Arab world closer were part of the driving force behind the Alexandria Conference which Presidents Mubarak and Assad and King Fahd attended in January 1995. Another motivating factor was Egyptian fears that the new Republican Congress in the United States would dry up vital US aid. The Conference was not aimed against the peace process, but the question of better Arab control over the process did arise. A few weeks after the Alexandria Conference, another conference was held in Cairo which revealed a new and extraordinary grouping in the Middle East. President Mubarak, King Hussein, Prime Minister Rabin and Chairman Arafat met to discuss ways to overcome difficulties and to consolidate the peace process.

Israel is concerned about the dwindling time left in which to reach and consolidate an agreement. Elections must be held in Israel by spring 1996. Rabin's popularity has been steadily eroding as it becomes clear that the agreements already reached have not brought security with them, and the elections could well be won by right-wing parties whose far more inflexible stand on territorial concessions would destroy any possibility of peace between Israel and Syria. Rabin has already said that he would hold a referendum should an agreement with Syria involve a withdrawal from the Golan Heights. It would be impossible for him to venture a massive withdrawal involving the dismantlement of dozens of settlements on the Golan, some dating back to 1967, with an election looming.

Exploding Hopes

In May 1994 the Israeli Defence Forces evacuated the Gaza Strip and Jericho area and Palestinian forces replaced them. For the first time, the PLO had succeeded, with Israeli consent, in establishing itself formally inside Palestine. Although the PLO looked on it as only a beginning, it was nevertheless a remarkable achievement.

The move had been preceded by two more agreements reached between Israel and the PLO, which are commonly combined and called the Gaza and Jericho Accord. The first was signed on 29 April 1994 in Paris and dealt with future economic relations. This agreement was followed by one signed in Cairo on 4 May which covered in detail the modalities of the withdrawals from both areas, the responsibilities of both parties, the transfer of civil powers, various security arrangements, the release of prisoners, and the movement of Palestinians between the Gaza Strip and Jericho, as well as arrangements at border crossings. At the same time, the Palestinians held talks with other countries on transferring promised funds, and the provisions necessary to acquire them.

In the first instance, Israel agreed to allow 9,000 Palestinian police and military to enter the evacuated territories. They were gathered from different parts of the Arab world – Egypt and Jordan, but also from Iraq, Sudan, Yemen and a number of North African countries. The Palestinians would like another 2,000 in their police force; in the meantime a number of police have been recruited from among the inhabitants of the 'liberated' Territories. The Palestinian force had little previous police training to prepare it for its main duties, and received military training separately in the various countries of origin. The lack of police, and riot control, training was immediately apparent when they quickly resorted to firearms causing 13 civilian deaths during civil disturbances organised by *Hamas* in the Gaza Strip.

Israeli Fears

Cooperation between the Palestinian and Israeli forces has been quite high, both in joint patrols on main arteries and in intelligence exchanges. But everything else has created serious and difficult problems. The most serious hurdle from the Israeli point of view is the persistence of attacks by Palestinian extremists. From the signing of the agreement on the White House lawn in September 1993 to the end of January 1995 at least 120 Israelis, many inside Israel, have been killed by knife, gun or bomb. Suicide bombers from *Hamas* and Islamic *Jihad* caused the major casualties; no operations appeared to be the work of the *Fatah* movement. On 29 October 1994 a passenger on a bus in a central Tel Aviv thoroughfare exploded a bomb he was carrying, killing himself and 21 other passengers. On 11 November a man on a bicycle killed himself and three Israeli soldiers when he rode into a checkpoint and triggered an explosion. In January 1995 at Beit Lid, two suicide bombers took over 20 Israelis with them when they destroyed themselves outside a pub frequented by Israeli soldiers.

Opinion polls have tracked a significant and consistent loss of Israeli support for the peace process and the Rabin government as a result of this wave of violence. To many Israelis it appears that peace has offered no security, and thus should not be called peace. Even the President of Israel, Ezer Weizmann, a man known for his early and strong support for peace negotiations with the PLO, has uncharacteristically, and probably unconstitutionally, called for a suspension of the talks.

Israel does not expect the Palestinians to wipe out terrorism, a goal that it had itself failed to achieve when it was in control of the Territories. What Israel does expect are all-out efforts. Israel accuses the Palestinian entity of shirking the commitments that it made in the Oslo agreement and accompanying letters. According to these agreements, murderers seeking refuge in the Gaza Strip must be handed over to Israeli jurisdiction; persons without a licence to carry weapons must be disarmed (the agreement provides for only the Palestinian police to carry arms – ownership of revolvers requires a licence); and, of great symbolic importance, the commitment to annul those clauses in the Palestinian charter calling for the destruction of the State of Israel must be implemented.

In Israeli eyes, the persistence of terrorism and the failure of the Palestinian entity to deal forthrightly with it is the main reason for the sluggish implementation of various clauses in the agreement for the interim stage. While Israel has handed over part of its civil powers to the Palestinians on the West Bank, it is unwilling to redeploy its forces there. The Israeli security forces have advised the government that while terrorism lasts they cannot be responsible for the security of Israelis living and working on the West Bank if their forces were moved from their present positions. There is deep concern that the territories will provide a base and asylum for terrorists. As those perpetrating the terrorism hoped, the uncontrolled outbreak of terrorist acts created a deadlock in the implementation of the second, interim stage of the agreement.

Palestinian Charges

The Palestinians also have many justified complaints. They say that in its reluctance to redeploy its forces before the elections, Israel has been holding up the elections to the Palestinian council. They complain that Israel has not fulfilled its part in arranging for 'safe passage' between the Gaza Strip and the Jericho area. They demand that Israel release all prisoners as undertaken in the agreement. They argue that the Israeli government is undercutting the agreement by continuing to build in the region of Jerusalem, and by allowing Israeli settlers to seize land that belongs to Palestinians and building more dwellings there.

The Palestinians are deeply concerned about the periodically imposed blockades. Tens of thousands of Palestinians are barred from reaching their work places in Israel as a result of the measures imposed by Israel after a serious terrorist act, not as punishment, but in an effort to prevent another such act. The uproar in Israel after the Beit Lid bombings has led the government to consider sealing off the Territories from Israel with high fences (some carrying a mild electric charge). The idea of maintaining a complete separation between the two communities has been given new impetus by the realisation of the near impossibility of preventing suicide bombers from fulfilling their desire for a glorious death.

The blockade undoubtedly hits hard at the many Palestinians who earn their living in Israel. Many employees in Israel have begun to avoid hiring Palestinians even though their wages are relatively low, their skills relatively high, and many speak Hebrew. They are being replaced, with government help, by an increasing number of imported workers (in early February the number had reached 70,000 from Europe and Asia), although in many cases these workers lack the skills and language ability of the Palestinians.

The damaging effects of the terrorism, and Israel's efforts to cope with it, are serious. Not only has there been a loss of overall income for the Palestinians, but the climate created by terrorism has deterred investment in the Territories. The state of the economy in the Gaza Strip and Jericho has deteriorated since the withdrawal of the IDF and the transfer of power to the Palestinian authority. There has been a sharp rise in unemployment with a concomitant serious decline in the standard of living. Countries which had promised an infusion of cash have held back their contributions on the basis that they have had difficulty convincing Arafat to create the transparency necessary for foreign investment. Although the Palestinians have called on Western donors to fulfil their commitments, they are also angry at Arab countries, especially the rich ones, for failing to live up to their promises.

Negotiations between Israel and the Palestinian Authority have continued, either under the auspices of President Mubarak in Cairo, or in Washington with impetus from President Clinton. But resolution of the various issues in dispute remains frozen. Israel, while pressing Arafat to participate in a joint effort to crush terrorism, has proposed two possible methods of advancing to the next stage. One is for the IDF to leave the towns and areas with Palestinian populations while elections are held, and to return after the elections.

The alternative is to postpone elections pending comprehensive negotiations on all subjects concerning the interim stage, including a more extensive military redeployment of Israeli forces, agreeing the powers of the Palestinian council, and transferring all remaining civil powers. Implementation of the agreements reached would not begin until the close of the negotiations, but Israel in anticipation would build roads circumve ng Arab towns to enable Israeli inhabitants to move separately in these areas when the extent of Palestinian control is negotiated. Arafat has apparently reluctantly accepted this second option, if only because Israel would not otherwise consent to a faster withdrawal and he would rather wait before arranging elections that could only be held with great difficulty if the Israeli troops did not withdraw or withdrew only with the assurance that they would be back immediately after the elections.

Overcoming Extremism

The fundamentalists on both sides, the orthodox Jewish settlers and the more radical Arab groups, can be expected to continue their campaigns to scupper the peace agreements and further negotiations. They have had a devastating effect thus far. Yet the logic of developments should point to a more positive conclusion to the peace efforts that have already been made. The majority of Israelis and Palestinians are more moderate in their beliefs, and want to live in peace, even if it is more a peace of separation than of integration.

Both the US and Egypt have recognised the need to become more engaged and are doing so with greater consistency and coherence. In mid-March, US Secretary of State Warren Christopher made his eleventh trip in two years to the region. At the end of the eight-day visit, which included talks in Cairo, Tel Aviv and Damascus, he announced that Israel and Syria would reopen negotiations in Washington, with military chiefs joining the diplomats in the bargaining. The United States intends to play a more active role in the negotiations; Christopher's senior aide, Dennis Ross, will participate in the discussions and 'make suggestions'. While Assad's willingness for talks to resume does not guarantee a positive outcome, he is under some pressure from the fact that elections due in 1966 in both the US and Israel are likely to bring in governments less friendly to the present peace process than those now in office.

The pressures on the other leading figures are also strong. Unless he can solidify further agreement with Israel and thus have a mini-state to lead, Arafat's political life is over. Rabin and the Labour Party face elections in mid-1996; only further advances towards peace will provide him with the basis for a successful campaign. Both leaders, thus, will need to compromise to assure their own survival. In talks in mid-March 1995, Israel and the PLO set a 1 July target date for an agreement on expanding Palestinian self-rule. Between such a statement and execution, of course, falls many a shadow. But the meetings at least have begun to break the long deadlock that the peace talks had reached, and both sides can be expected to try to widen the areas of agreement between them.

Three Threats from Radical Islam

Regimes in Algeria, Egypt and Saudi Arabia all face an Islamist challenge. Algeria is fighting for survival. Egypt, despite its low-intensity conflict with Islamic militants, is not. Nor, more emphatically, is Saudi Arabia. But Saudi Arabia and Egypt both fear that a collapse of the military regime in Algeria will embolden their own Islamist opponents. All three regimes share common perceptions of the Islamist 'threat'. These were embodied in an anti-terrorist 'Code of Conduct' (of which Algeria and Egypt were leading sponsors) adopted at the Casablanca Islamic summit in December 1994. The same theme dominated the Arab Ministers of Interior conference in Tunis the following January.

Disintegration in Algeria

The Algerian regime has a three-pronged strategy to restore stability and maintain power. Military action and economic reform have been two key elements for some time. Following the breakdown of dialogue with the *Front Islamique de Salut* (FIS) in October 1994, the regime has added the promise of presidential elections before the end of 1995. This third prong is unlikely to be any more effective in placating the opposition than the other two have been, however.

Open-ended Violence
Hardliners who dominate the military regime remain convinced that they can defeat the Islamist opposition by military and security means without resorting to political compromise. Repression has been the corner-stone of

the regime's counter-insurgency effort and has been conducted with heightened intensity since autumn 1994 in an all-out war to eradicate the armed opposition. But government forces, which are bearing the brunt of the insurgency, are increasingly stretched. The pressures of mounting violence, death threats and other forms of intimidation by their opponents will have had an impact on morale. There have been reports of desertions and defections, particularly among the lower ranks.

The insurgents have proved resolute, resourceful and resilient. Their attacks on police and military facilities have given them access to weapons and equipment which has helped to underpin their continuing offensive capability. The December hijack of an Air France aircraft by the *Groupe Islamique Armée* (GIA) exposed how vulnerable Algerian security was to Islamist penetration. And the suicide car-bomb attack on the Algiers central police station on 30 January, which killed 42 people and wounded several hundred others, represented a new and disturbing tactic which will almost certainly be repeated. According to government figures published in early 1995, there were nearly 3,000 acts of sabotage in Algeria during 1994 (causing damage worth $1.3 billion). This campaign continued unabated throughout February 1995, with daily reports of damage to schools, factories, railway bridges and public utilities. The assassination of journalists, intellectuals and local government officials has also continued. Government forces have more than matched such atrocities. Their brutal suppression of a mutiny in an Algiers prison in late February resulted in the death of at least 95 prisoners.

Searching for Political Cover

The regime's efforts to bolster its legitimacy began with the creation of an appointed consultative body, the Transitional National Council, in May 1994. This was boycotted by nearly all the main political parties as a meaningless rubber-stamp. Later in the year, President Liamine Zeroual initiated a series of talks with FIS leaders, releasing five and moving the two principal leaders, Abassi Madani and Ali Belhadj, from prison to house arrest. Zeroual, the figurehead of the generals who appointed him, had little room for manoeuvre in his attempts to persuade the FIS to condemn violence. Their short-lived dialogue appears to have been a ploy by the government to divide the Islamist opposition.

In his speech on 31 October 1994 announcing the breakdown of the dialogue (for which he blamed the FIS), Zeroual sought to appeal directly to the Algerian people, over the heads of all opposition parties, with the promise of presidential elections before the end of 1995. The regime evidently sees such elections, from which the FIS are to be excluded, as a means of rallying the support of uncommitted Algerians and forging a political centre aligned with the government from parties which agree to participate.

Zeroual has not declared himself as a candidate, but is viewed as the regime's obvious choice. In December, during his first trip outside Algiers since assuming office in early 1994, Zeroual visited the south where he promised voters increased aid for development. He has also endorsed Berber demands for the official recognition of their language in a bid for their support. Meanwhile, a seven-man election reform commission headed by Prime Minister Mokdad Sifi has been appointed to consider changes to the electoral law. In December, Sifi claimed that before conducting the elections the government would rebuild 178 city halls destroyed by insurgents, would replace mayors and other assassinated officials, and would deploy 45,000 election monitors.

Even if the regime could find the people to fill these posts it is hard to see how its electoral strategy can succeed in the current climate of violence and intimidation. The major opposition parties have reacted unfavourably, and on 13 January 1995 in Rome, following meetings under the aegis of the St Egidio Catholic community, they unveiled their own blueprint for a political solution to the country's crisis. The St Egidio conference was notable for bringing together representatives from all Algeria's main opposition parties – including the FIS – that had won 80% of the vote in the December 1991 elections which were later annulled by the government.

The resulting agreement renounced violence and supported the principles of democratic pluralism. It listed several preconditions for negotiation, including the release of FIS leaders and political prisoners and the restoration of the FIS's legal status. By presenting a united front on these demands, the opposition put the regime on the defensive and enhanced the FIS's international credibility. But the Rome agreement raises questions about the degree to which a consensus has been or could be achieved within the FIS and its associated militias in support of the moderate platform embodied in the agreement.

The regime angrily rejected it and seems intent on pursuing its own initiative to organise controlled presidential elections. But the Rome plan received cautious support from European Union (EU) foreign ministers on 23 January as the possible basis for ending a conflict in which over 30,000 Algerians and some 90 foreigners have been killed.

Economic Reform

France, which is Algeria's leading trading partner and provides it with bilateral aid of $1.1bn a year in the form of credits, has taken the lead in mobilising European and international economic assistance to the regime. In 1994, the International Monetary Fund (IMF) awarded Algeria $1bn and the Paris Club of creditors rescheduled loans worth $5bn. In 1995, Algeria will require $8bn in debt relief and new money to keep its economy afloat. It has already signed a letter of intent with the IMF which once approved would enable it to draw $500m a year over the next three years. A deal with the IMF will pave the way for financial aid from the World Bank and a second rescheduling of the debt owed to the Paris Club.

According to the IMF, Algeria has scrupulously adhered to IMF prescriptions by cutting its deficit, devaluing its currency and freeing prices. Western governments recognise that halting aid to Algeria could fuel further chaos. They also recognise that the regime's policy of 'eradication' has failed. Yet the West has shown no inclination to translate its increasing rhetorical support for a political solution into hard conditionality by linking further aid and debt relief to the regime's willingness to negotiate. This reflects European, especially French, fears that an Islamist takeover will result in a wave of Algerian immigrants into Europe and would encourage similar activities in neighbouring states.

The rationale for continued aid with only economic strings is that reform is working and will ultimately rob the Islamist opposition of their support among the poor and unemployed. But the reality on the ground presents a different picture. With high inflation the purchasing power of most Algerians has fallen by roughly 20% in the past year, enough to force many families to change their work and consumption patterns. Unemployment, estimated at 30% in 1994, is set to rise, particularly if government plans to privatise state-run enterprises are implemented. The deteriorating security situation is likely to endanger prospects for economic recovery by triggering a further exodus of foreign experts and derailing plans for expanding the country's vital hydrocarbon sector. Oil and gas revenues account for more than 95% of Algeria's hard-currency earnings.

The Outlook

The regime remains committed to its triple strategy of repression, presidential elections and economic reform. It has signalled its unwillingness to pursue a negotiated settlement by rejecting the Rome agreement. It will continue to survive so long as military cohesion does not disintegrate. Despite recent rumours of splits within the military establishment, its cohesion appears to be holding. But the regime's conviction that it can defeat the insurgents unconditionally has not been borne out by events. Although the Islamist armed groups are divided by factional, regional and ideological differences they enjoy an influence disproportionate to their numbers and resources. They, too, appear confident of victory. The prospect remains of protracted stalemate with more violence and bloodshed, and a gradual but steady deterioration of security.

Containment in Egypt

Unlike Algeria, the government in Egypt has gained the upper hand in its conflict with extremist groups operating under the banner of the hydra-headed *Al Jama'at Al Islamiya*. This has been achieved at the cost of considerable resources and a resort to heavy-handed and increasingly aggressive tactics – hostage-taking, collective punishment, extra-judicial killings – which have been strongly criticised by local and international human-rights organisations. They may even have rallied fresh recruits to the Islamist cause. But the overall result has been a marked decrease in violence in Upper Egypt, calm in Cairo and the revival of tourism.

Localised resistance continues, with a high level of bloodshed registered in the Governorate of Minya during January 1995. But for the present the disruptive potential of insurgent groups and their transnational support network seem to have been seriously weakened. Some 35 wanted 'terrorists' have been returned from abroad under cooperative security agreements signed with other states, notably Pakistan. The Minister of Interior has claimed that his job is 85% done and justifies his failure to eliminate domestic violence saying that Egyptian extremists continue to receive material support from sympathisers in Europe, Sudan and Iran.

The government has also taken initiatives in the media. Soap operas aimed at exposing the hypocrisy and opportunism of the *Jama'at*, the televised confessions of repentant extremists and interviews with religious leaders condemning violence as alien to Islamic teachings have helped to undermine the credibility of the militants. Moreover, in a society which is still governed in the Pharaonic style, Egyptians remain wary of the power of central authority. The regime can exploit this, especially in the country where over 40% of the population live and only the government has the resources to offer patronage and some prospect of upward mobility.

A Strategic Tilt

President Mubarak has always viewed the non-violent wing of Egypt's Islamic movement, embodied in the Muslim Brotherhood (MB), as the only significant challenge to his regime. For this reason he has refused to legalise it. The emergency laws in force since 1981 give him the power to act against it at will. Until recently the regime was content to play a cat and mouse game with the MB, but in 1994 there were signs of a more confrontational approach.

Since 1992 the MB has dominated the boards of Egypt's main professional bodies, including the Lawyers', Engineers' and Doctors' Syndicates. They have also been influential on the campus. These successes owe more to good organisation than to mass support. Despite the introduction of legislation in January 1993 to bring syndicate elections under tighter state control and thus prevent further MB gains, in 1994 the MB added the Agronomists' and Scientists' Syndicates to the list of those they control. The regime responded by rushing through legal changes in February 1995 which in effect place the control of electoral registers in the hands of judicial commissions appointed by the government.

Meanwhile, President Mubarak pointedly excluded the MB from the National Dialogue. This was finally held in June 1994, but was boycotted by the two largest legal opposition parties, the Wafd and Nasserists, and proved to be the charade that they had predicted. The regime also took direct action against the MB. Members of the MB-dominated Lawyers' Syndicate were arrested for distributing pamphlets, a technically illegal act but one which had been previously tolerated. Even the Supreme Guide, the MB's 84-year-old titular head, was questioned. Led by Mubarak, official criticism of the MB became increasingly strident. The President accused it of being directly implicated in 'terrorism' and of channelling funds from abroad to Islamic extremist groups.

In mid-December Adel Hussain, Secretary-General of the Labour Party, the main opposition ally of the MB, was arrested on a charge of alleged links with the extremist *Jama'at*. This was widely seen as part of a government campaign to put pressure on the Labour Party to break with the MB and thus deprive it of an electoral platform in the run-up to parliamentary elections in 1995, which the MB had declared they would contest. In October, the Liberal Party, which also has links with the MB, had been the target of government harassment.

In January 1995, 27 leading members of the MB, including the Assistant Secretary-General of the Doctors' Syndicate, a former member of parliament, were arrested on charges of plotting to overthrow the state. These arrests were accompanied by a press campaign linking the MB with Egypt's insurgent groups and their alleged state sponsors, Iran and Sudan. Although it is the largest opposition movement, the MB has neither the strength nor the will to mount a violent challenge to the regime. It has confined itself to accusing the regime of fabricating charges to intimidate it prior to national elections.

More of the Same

Mubarak has consistently shown throughout his 14-year presidency that he will do nothing to put at risk his ruling party's grip on power. Although he allowed the MB to run in past elections under the flag of legal secular parties, the lesson he has drawn from Algeria's experience is that Islamists and elections can be an explosive mix. His policy of discrediting and delegitimising the MB can be seen as an attempt to control the Islamist 'virus' at an early and still manageable stage. His gamble may work in the short term, but the MB's historical tendency to split under pressure into radical and moderate wings may come back to haunt him. Younger members of the MB, particularly university students and recent graduates, will be tempted to conclude that the peaceful path to change is futile. Significant MB defections to the extremists would exacerbate the government's problems and accelerate the cycle of violence and repression.

The Islamic trend and its violent fringe feed off the social and economic problems, which include high unemployment, lack of affordable housing and basic services, disparities of wealth and entrenched corruption (which even touch Mubarak's family and inner circle). These remain largely unaddressed. The regime's fear that the Islamists will capitalise on any socio-economic discontent arising from IMF prescriptions for economic reform has led the government to resist fully implementing these reforms, thus forgoing further debt-forgiveness from foreign creditors. The reduction and removal of subsidies has already increased pressures on the poor. In October plans to privatise state industries caused labour unrest and the threat of strike action which the regime summarily quashed.

In a keynote speech at the International Book Fair in Cairo in January 1995, Muhammad Hassanain Heikal, the veteran Egyptian journalist and commentator on Arab affairs, linked the country's socio-economic problems to its political stagnation. He spoke of an unprecedented degree of public alienation and despair which was fuelling a mood of anger and a yearning for change, of which armed violence was only one manifestation. He urged Mubarak to make the election year 1995 an occasion for radical political and administrative reform. This he saw as the key to solving the crisis. His call is unlikely to be heeded. Egypt's intelligentsia have freedom to criticise, but no means to change the system.

Saudi Arabia

The Al Saud came to power in alliance with the Wahhabi revivalist movement and their relationship with the religious establishment remains the basis of their legitimacy. Since the 1991 Gulf War the Al Saud have been confronted with a growing Islamist movement calling for sweeping changes in the country's political and social life. This may seem anomalous in a state which prides itself on ruling in strict accordance with Islamic law, but is not so strange given that the regime demonstrated during the Gulf War its willingness to invite 'infidel' forces to protect the 'land of the prophet'.

The Challenge

The nature of the challenge can be gauged from the substance of two petitions addressed to King Fahd by groups of religious scholars and intellectuals in May 1991 and September 1992. The first (with 52 signatories) called for the creation of an independent Consultative Council with the power to decide domestic and foreign policy, for equality before the law, for the redistribution of wealth, for the accountability of all officials and for stricter adherence to Islamic values. The second petition (with 107 signatories) elaborated on these demands, drawing attention to mismanagement and corruption in public life and making detailed recommendations for remedial action. These envisaged a major role for the clergy in decision-making and the control of public expenditure. They also included calls for the exclusion of foreign cultural influences and the reorientation of foreign and defence policy by cutting ties with the West and non-Islamist Arab regimes.

Many of the signatories were men of religion – prayer leaders, preachers, jurists and teachers – from the Najd, the cradle of the Wahhabi movement and the traditional power base of the Al Saud. Several of them, including two charismatic preachers (Salman Al Auda and Safar Al Hawali) were to pose a more direct challenge to government authority during the next two years. In essence, this grand remonstrance demanded basic structural changes involving a major transfer of power from the Royal family to a self-appointed elite of religious puritans.

It was denounced as misguided and divisive by all but seven members of the Supreme Council of *Ulema* (the latter were 'retired' and replaced by younger and more liberal blood). Significantly, the regime's long-promised Consultative Council (*Majlis ash-Shoura*), finally inaugurated in late December 1993 as the government's cautious response to growing pressure for political participation, was to exclude the Al Saud's Islamist critics. The majority of its 60 appointed members were Western-educated technocrats and academics, with only a handful of religious notables, selected for their loyalty.

The Response

The regime's initial response to the Islamist challenge was characterised by a blend of indecision and appeasement. It was in the King's nature to avoid confrontation. The conspicuous consumption of the Royal family, against a backdrop of budgetary pressures, made the regime vulnerable to criticism. And it was not easy to rebut criticism of the regime's Islamic credentials endorsed by dissident members of the religious hierarchy amid calls for stricter Islamic discipline. However, in an address to religious scholars in December 1992 the King warned religious activists to stop disseminating anti-government propaganda in mosque sermons, pamphlets and cassettes, which he described as a new and unfamiliar phenomenon. He stressed that the Kingdom adhered fully to Islamic tenets. The authorities were open to any complaints or advice that citizens might wish to offer privately. But the pulpit should be confined to spiritual matters. The government would no long turn a blind eye to those who exceeded these 'limits'.

In May 1993 five signatories of the September 1992 petition announced the formation of the Committee for the Defence of Legitimate Rights (CDLR) to lobby for civil liberties and to monitor human-rights abuses in the Kingdom. Many Western observers welcomed the CDLR as a new recruit to the world's human-rights movement until it became clear that the CDLR's definition of human rights, theocratically inspired, fell far short of universal standards. The Saudis moved quickly to declare the CDLR illegal and to disband it. Its spokesman, Muhammad Masa'ri, a

Western-educated physics professor, after a period of detention, fled to the UK in April 1994.

Since then, from offices in London, the CDLR has waged a fax war against the regime, concentrating on tales of princely corruption, repression and mismanagement. The aim is clearly to wash as much dirty linen in public as possible. The CDLR seems to have access to a ready supply of material from correspondents across the Kingdom and modern information technology makes it easy to magnify local discontents. The CDLR is playing opportunistically to a disparate constituency of dissent. But their zeal to discredit the system by vilifying senior princes seems likely to alienate the technocratic elite.

Crackdown

September 1994 marked a turning-point in the regime's confrontation with the Islamist opposition, with the arrest of the popular preachers Auda and Hawali who had defied warnings from the Senior Council of *Ulema* to desist from mixing religion and politics. Both men had taken the lead in condemning the presence of foreign forces on Saudi soil during the Gulf crisis. Both had attacked government corruption and aspects of foreign policy, including Saudi support for the Middle East peace process and the regime's close ties with the West. Cassettes of their sermons had circulated widely, especially on the university campus and among the disaffected younger generation.

Their detention sparked demonstrations in Qasim and the arrest of over a hundred sympathisers, most of whom were later released. A statement by the normally tight-lipped Ministry of the Interior giving an account of these events was precipitated by exaggerated reports of unrest issued by the CDLR and carried by the international media. In October the Ministry issued a stern warning that it would take deterrent measures against those whose actions ran counter to Islam or undermined security. This reflected government fears that religious activism might become a vehicle for broader protests. Meanwhile many prominent *Ulema* rallied to support the government in sermons and press interviews, stressing the religious duty of the faithful to obey the authorities.

Pre-emption

The regime's efforts to introduce administrative measures to control radical preaching and the use of mosques as platforms for opposing government policies had already begun in 1993 with the creation of a new Ministry of Islamic Affairs whose writ included responsibility for the country's Islamic universities. Government machinery for enforcing religious discipline was further strengthened in October 1994 with the creation of a 14-member *Da'wa* (religious propagation) Council headed by the Minister of Islamic Affairs. Its role was to supervise all matters relating to mosques and sermons, to vet prayer leaders and preachers and to oversee educational programmes designed to 'protect youth from radical ideas'. This initiative reflected the religious hierarchy's failure to bring dissident activists to heel, but did not directly challenge the senior *Ulema* in matters of theology, nor their formal independence.

The same month another body, the Supreme Council for Islamic Affairs, was formed under the chairmanship of Prince Sultan, Saudi Minister of Defence. Its membership includes the Ministers of the Interior, Higher Education, Finance, Justice, Foreign Affairs and the Secretary-General of the Muslim World League, the main channel of Saudi assistance to Islamic causes. The role of this heavyweight body is to coordinate all Saudi aid to Muslim minorities and Islamic societies abroad to ensure that it does not reach what Prince Sultan defined as politically active groups operating under the mantle of Islam.

Saudi Arabia has often been accused of funding dissident Islamic groups in other states or of failing to prevent private Saudi money from reaching such groups. In 1993 the government introduced measures to ensure that all private/mosque donations to Islamic causes were disbursed through official bodies. And in April 1994, in response to Egyptian pressure, the Saudis took the unusual step of stripping a member of one of its wealthiest merchant families, Usama Bin Ladin, of his Saudi citizenship for reportedly funding Egyptian and other extremist groups.

Economic Pressures

The government's efforts to manage religious and political dissent need to be seen in the context of the Kingdom's financial problems and its transition from a political economy of abundance to one of relative scarcity. For some years, Saudi Arabia has been living beyond its means, running large budget and current-account deficits. This position has been complicated by the enormous cost of assisting Iraq during its war with Iran ($25bn), the cost to the Saudis of the Gulf War ($55bn) and falling oil prices. The government has avoided fiscal discipline by drawing down its reserves (reduced from $120bn to $15bn), borrowing and delaying payments. In 1994 government spending was cut by 20%. The 1995 budget foreshadows further cuts in public spending and increases in the price of utilities and services, including electricity, petrol, telephone, water and domestic air travel. The King used the opening of the annual session of the Consultative Council in December 1994 to announce these price rises, presenting them, disingenuously, as a temporary expedient. His chosen platform suggests a wish to spread responsibility for inflicting even modest economic pain.

Outlook

The Saudis have long performed a skilful balancing act between the requirements of a fast-developing oil-fired economy and religious tradition.

They have won loyalty and neutralised dissent with subsidies and a cradle-to-the-grave welfare state. Since the Gulf War a politically active segment of Saudi society has emerged from the ranks of religious activists, pressing for cleaner, more accountable government, stricter public morality and a more active support of Islamist causes, as in Bosnia and Palestine. Islam affords the only ideological and organisational basis on which self-proclaimed reformers can challenge government policy. The Al Saud are hostage to their own claim of Islamic legitimacy.

The Islamist opposition draws on the support of a rapidly growing younger generation (almost 70% of the population are under 25) whose religiously based education, increasingly inadequate for the needs of a modern society, threatens to marginalise them. Many are unemployed and resent their declining economic prospects, royal corruption and Saudi dependence on the West. In the absence of other outlets for dissent, the mosque and campus have become centres for political debate. Cuts in subsidies are likely to seriously aggravate these frustrations, especially if the Royal family are not seen to share the burden. Financial pressures diminish the government's freedom of action, and high population growth (nearly 4%) is adding to the problem of resource allocation. Nevertheless, the Al Saud has responded with firmness, taken care to keep the religious establishment on its side and taken unprecedented steps to improve the public presentation of its policies. It must keep its nerve and show greater sensitivity – not least a willingness to tighten its own belt.

Iraq: Still Squirming under Sanctions

United Nations (UN) sanctions against the rogue Iraqi regime remained in place during the past year despite the strenuous efforts of Iraq's government to have them eased or lifted. Saddam Hussein may have taken heart from increasing signs of discord among the permanent members of the UN Security Council (UNSC) on the issue, but this discord failed to create sufficient momentum to free Iraq from the economic embargo. During 1994, Iraq and the sanctions issue were caught up in several timetables.

The first was the bi-monthly formal renewal of the UN sanctions regime. Despite its general *pro forma* nature, this review began to encourage the development of political pressures within the UNSC that argued for the easing of sanctions. The Iraqi government obviously hoped that this would accelerate and thus overtake the threatening domestic pressures of popular economic destitution and clannish disaffection within the ruling establishment. Meanwhile, a similar sense of a race against time shaped developments within the northern Kurdish zone, as rival Kurdish parties fought each other against a background of growing uncertainty.

The Sanctions Argument

The Iraqi government had hoped that its renewed cooperation with the UN Special Commission on Disarmament (UNSCOM) would lead to the easing of sanctions, in particular the lifting of the embargo on oil sales. In February 1994, Rolf Ekeus, the UN special representative and head of the disarmament commission, felt confident enough to begin talks with Iraq about installing long-term monitoring equipment to track Iraqi weapons production and testing sites.

Iraq and others saw the installation of such equipment as heralding the beginning of the end of the sanctions regime, marking the start of a defined probationary period during which Iraqi compliance would be monitored. If it were seen to comply fully during this fixed period the Iraqi government believed the sanctions regime would be eased. It was no doubt encouraged in this by UNSCOM's verdict during the same month that it could find no evidence of chemical weapons being used by Iraqi forces in the southern marshlands of Iraq during 1993.

In this atmosphere the first indications of disagreement about the wisdom of continuing the sanctions regime began to emerge within the UN Security Council. In mid-March the sanctions regime came up for its bi-monthly review. Hitherto, the renewal of sanctions had been a routine affair. This time it became clear that differences of opinion were beginning to appear among the permanent members. The UNSC agreed on its re-newal, but only after some days of intense discussion.

France wanted to include a statement in the renewal order praising Iraq for its compliance with the UN and was supported by Russia and China. In addition, Russia wanted a definite indication of when sanctions would end. These suggestions were rejected by the United States and the United Kingdom. In particular, the US stated that Iraq must comply with all other UN resolutions (principally UNSC Resolution 688), not simply those relating to disarmament. The US and UK held that the embargo could not be lifted until Iraq both formally recognised the independence of Kuwait and the recently demarcated Iraqi–Kuwaiti border, and demon-strated respect for human rights within Iraq itself.

These differences of opinion in the UNSC became more pronounced during the year. France and Russia argued for easing the sanctions regime as Iraqi compliance with the disarmament resolution became clearer. Naturally, the Iraqi government favoured this interpretation of compli-ance with what it called 'all relevant UN resolutions' (principally, UNSC Resolution 687) and did its utmost to cultivate France and Russia in the hope that they would create enough pressure within the Security Council to have the embargo lifted.

Neither France nor Russia made much headway in the face of Ameri-can and British opposition. Nor was their case greatly strengthened by the knowledge that they were not disinterested parties. In the first place, France and Russia were two of Iraq's major creditors. Due in large part to arms sales during the 1980s, Iraq owed France $6–9bn and Russia $10–16bn – debts which could clearly not be discharged until Iraqi oil revenues began flowing again. Second, French and Russian state concerns and private companies had entered into negotiations with the Iraqi government for lucrative contracts that would come into force as soon as the economic sanctions were lifted.

At the same time, Baghdad's dismal human-rights record was being chronicled by Max van der Stoel, the UN special rapporteur on human rights. This helped to strengthen the case of the US and the UK. Yet within both countries there was some public unease about justifying the sanctions regime with reference to human-rights abuses when the sanctions were apparently having a harmful effect on the well-being of many Iraqis. The Iraqi government was careful to draw international attention to certain aspects of this economic situation in the hope of increasing the pressure.

The Continuing Threat

To the same end, Iraq complied fully with UNSCOM as it began to install the extensive weapon-monitoring system in July. But in September, as the time approached for the system to become operational, the Iraqi government threatened to 'reconsider' its attitude to the UN, and its compliance, if sanctions were not lifted as soon as the monitoring began officially. Ekeus' reply that this could not even be considered before a six-month period of proven compliance had elapsed enraged the Iraqi government. It began to talk even more threateningly of non-cooperation and of unspecified consequences if sanctions remained in place.

In the first week of October this threatening language took on a new and apparently more dangerous form when the Iraqis began to move 60,000 of their troops towards the Kuwaiti border. Given its unprovoked attack on Kuwait in 1990, the US felt that it could take no risks, even if the Iraqi government was indulging in an elaborate game of bluff. 18,000 US troops were transported to Kuwait and US air-strike capabilities in the area were reinforced. At the same time, the US warned the Iraqi government explicitly that if any move were made against Kuwait, Baghdad and the Iraqi regime would be targeted.

Iraq's manoeuvres were self-defeating. On the one hand, UNSCOM's positive report on Iraqi compliance with the disarmament programme was about to be released and would have supported the arguments of those members of the Security Council in favour of easing sanctions. On the other hand, the move implied desperation on the part of the Iraqi government, suggesting that the regime must be cracking under the strain of the sanctions.

Indeed, a number of developments appeared to have borne this out in the months preceding the Kuwaiti border crisis. In May, Saddam Hussein had blamed the dire economic plight of his people at least partially on his Prime Minister and the Minister of Agriculture, and dismissed the two from office. He himself assumed the office of Prime Minister, but this was soon followed by a further collapse of the Iraqi dinar. By September, he was forced to cut subsidies to a range of goods and to cut the rations of basic foodstuffs.

Saddam accompanied these moves with pay rises for the police and the armed forces. They were needed more than ever to administer the draconian punishments of amputation and branding introduced in June for those convicted of theft and desertion from the armed forces, as well as to suppress the numerous outbreaks of public disorder. At the same time, increasingly frequent reports of bomb explosions in Baghdad reinforced the impression of a regime under siege. Given these conditions, the execution during the summer of a significant number of officers from one of the major Sunni Arab military clans – the al-Duris – was an ominous development. It tended to indicate that suspicion and dissent were gradually creeping closer to the heart of power.

At one and the same time, therefore, Iraq's threatening posture had undermined the position of its potential allies on the Security Council, whilst convincing the US and the UK that sanctions might be having the desired – if unspoken – effect of bringing down Saddam Hussein's regime. The outcome was a unanimous Security Council vote condemning the Iraqi actions and a new UNSC resolution threatening air strikes against Iraqi Republican Guard Divisions if they moved south of the 32nd parallel in the future. The US underlined the seriousness of this threat by stationing more planes in Kuwait, and pre-positioning armour in Kuwait and Qatar.

The US and the UK also took this opportunity to spell out the conditions that Iraq would have to fulfil before sanctions could be lifted. In addition to the known question of compliance with disarmament requirements, these conditions were Iraqi recognition of the sovereignty of Kuwait and of the newly demarcated border, Iraqi cooperation in efforts to repatriate the hundreds of Kuwaitis missing since the occupation, as well as to recover plundered Kuwaiti property, and a marked improvement in Iraq's human-rights record.

These conditions gave Russia the opportunity to play an active mediation role. After intensive discussions during October, the Russian diplomatic effort produced an Iraqi commitment to recognise Kuwait as an independent sovereign state. This was formally announced by the Iraqi government in November, prior to the UN debate on the renewal of sanctions. Accordingly, both Russia and China called on the UN to start easing sanctions. But neither could prevail against the opposition of the US which refused to tie itself to any such timetable. At the same time the US produced evidence suggesting that the Iraqi government was not as short of funds as it claimed. Iraqi oil had been reaching the world market either overland by truck through Turkey, or by coastal shipping through Iran. The volume was not great (estimated at about 100,000 barrels a day), but it was providing a steady and useful source of foreign exchange for the regime.

The Iraqi government maintained its diplomatic offensive to ensure that each time the question of sanctions renewal came up for discussion at the Security Council those who called for renewal would come under increasing pressure. In January 1995 the French Foreign Minister, Alain Juppé, met Tariq Aziz, the Iraqi Prime Minister, and announced the imminent opening of a French diplomatic 'interests section' in Baghdad. This visible split in the ranks of the former Gulf War coalition took some by surprise, but did not prevent the renewal of the sanctions regime in mid-January. Nevertheless, the Iraqi government pressed on, turning its attention to the commercial sector by encouraging a series of visits by European businesses and by international oil companies in the month leading up to the March debate on the renewal of sanctions. This strategy was weakened, however, by indications that Iraq had been deceiving UNSCOM on both its missile capability and on its biological weapons programme, as well as by a further damning UN report in February on the human-rights record of the Iraqi government.

Having perhaps learnt its lesson from the mistakes it made in October, the Iraqi government sought to create a very different atmosphere in the run-up to the presentation of UNSCOM's April report on Iraqi compliance with the disarmament programme, but to no avail. In mid-March 1995 the UNSC voted to maintain sanctions. Iraq will thus be obliged to cope with their continuing damaging internal effects as the resources available to the regime diminish ever faster.

The Kurds at Loggerheads

Saddam Hussein may dream that sanctions will be lifted before their corrosive internal effects bring about the demise of his regime. For the Kurdish parties in the northern zone, however, this is their nightmare. Insecurity and fear about the future contributed to the increasingly bitter conflict between the Kurdish Democratic Party (KDP) and the Popular Union of Kurdistan (PUK).

These tensions erupted in May into violent conflict which left some 600 dead by the beginning of June. A cease-fire was arranged, but it broke down in August when renewed clashes between the KDP and the Kurdish Islamic Movement, on one side, and the PUK on the other, caused further heavy casualties. Tensions remained high and were not defused even when the leaders of the two main parties signed a peace agreement in November. By late December fighting had erupted once again causing hundreds of casualties. It subsided briefly in January 1995, but flared up soon after.

The conflict is due in part to the nature and constituencies of the two parties, as well as to their troubled histories. But it is also due to the struggle for power, territory and resources in the northern zone. The fiction of the coalition government formed after the 1992 elections became ever more obvious since effective leadership remained with the parties themselves and the militias at their command. These were concerned to extend their power and their revenue-raising capacities over as wide an area as possible.

It was inevitable, therefore, that the two parties should have come into conflict when they began to compete for the same territory or to claim the right to the same sources of finance. Mindful also of the extraordinary staying power of Saddam Hussein and of the pressures building up to rehabilitate Iraq under his leadership, both parties may have been entering into negotiations with Baghdad to ensure that they would not be at a disadvantage should the authority of central government be reasserted over the north once again.

For the Iraqi National Congress (INC) – the coalition of disparate Iraqi opposition forces – these were serious developments. On the one hand, the sight of the Kurdish parties at war with each other did not enhance the credibility of the INC, since the Kurdish parties are themselves important members of this body. The INC has been trying hard to convince outside powers – chiefly the US – that it is a credible alternative to Saddam Hussein's regime and this internecine dispute did nothing to enhance its credibility.

On the other hand, the breakdown of order in the Kurdish northern zone gave Saddam Hussein greater leverage there. It may also have reinforced his claim that the factions of Iraqi society needed to be ruled by someone such as himself in order to keep the peace between them – hence his offer to mediate between the parties in January. It was therefore not surprising that the INC should have thrown itself so enthusiastically into the thankless task of trying first to separate the warring Kurdish parties and then to mediate between them. It claimed some initial success in both areas, but in the longer term this process will depend upon forces beyond the control of the INC itself.

Time is Running Out

During the past year, therefore, the calculations of all parties, whether in government or in opposition, have been dominated by the prospects of the lifting of sanctions and the possible reintegration of Iraq into the international community. The Iraqi government has been encouraged by the possibility that the timetable of international pressure, commercial and diplomatic, has been working in its favour. This has led to debate within the Security Council and a clear division among its permanent members over the question of easing the sanctions regime.

Iraq fears that the processes of internal dissent and fragmentation will work at a faster rate than that of the momentum for lifting sanctions, threatening to bring down the regime before the oil revenues can start flowing again. The fear of the INC is the opposite: that the sanctions will be eased whilst Saddam Hussein is still in power, strengthening his credibility and his regime and giving him the opportunity to reassert his control over the whole of Iraq.

Conflict in Yemen

During the past year the tensions inherent in the 1990 unification of North and South Yemen exploded into open warfare. The coming together in May 1990 of the Yemen Arab Republic (YAR) and the People's Democratic Republic of Yemen (PDRY) to form the Republic of Yemen was not as wholehearted a unification as was portrayed at the time. It was true that the border between the countries had been abolished, that the senior politicians of the two states had become joint office holders in the new Republic, and that a parliament was established with representation from both north and south. But the political organisations and the governing apparatus of both states, including their security forces, remained intact and separate, answerable to their erstwhile leaders. In fact, if not in name, two states continued to exist in Yemen.

The deteriorating economic situation of the country as a whole and the discovery of oil in parts of the south merely added to the mutual suspicion which had been growing between the dominant parties in north and south. The antagonism between them had been characterised by an increasingly violent political life in which members of the southern Yemen Socialist Party were frequent targets of assassination and intimidation. The terrorism reinforced the fears of the southern leaders that their power base among the numerically smaller population of the south (roughly 20% of the total population of the Republic of Yemen) was under threat. These considerations led the southern leader, Ali Salim al-Baidh, to leave the capital, Sana'a, in 1993 and to retreat in protest to the greater security of the former southern capital Aden.

In February 1994 a meeting between al-Baidh and the northern leader, Ali Abdullah Salih, was organised through Jordanian mediation in Amman. This led to the signing of the Amman Accord, a document of 'national reconciliation' intended to reassure both sides and to encourage them to cooperate with confidence once again. In practice, the accord failed to create the requisite trust. It was signed against a background of increasingly frequent clashes between units of the armed forces of the two sides which had been stationed for some years in each other's territory. After the signing these clashes continued, leading to increasing numbers of casualties. Such was their frequency and scale that they seemed no longer to be simply random outbursts. Rather, both sides seemed to be jostling for tactical advantage as part of larger strategies.

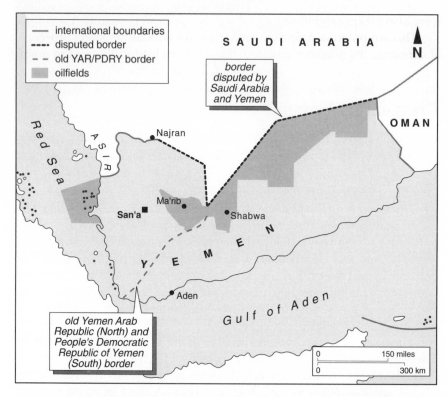

By April, assassinations or assassination attempts against northern and southern officials were becoming more frequent. Meanwhile, the fighting escalated to the bombardment of civilian areas in north and south beyond the confines of the military bases of the army units directly involved. In the first week of May the outbreak of general warfare between the two sides was signalled by air raids launched against Sana'a and Aden. At this stage there were few public declarations of intent by the northern and southern leaders. Nevertheless, in retrospect, it appears that the northern leadership had decided that it was time for a show-down with the southerners implementing the second stage of unification, but one based now on terms dictated by the dominant partner to the north.

As far as the southern leadership was concerned, the initial use of force seems to have been intended as a warning that the southerners would fight to preserve their relative autonomy. The hope presumably was to deter the northern leadership from seeking to impose its will by force. Once it became apparent that the northern leadership was undeterred because it was confident of its military superiority, the southerners chose the fourth anniversary of unification to declare the secession of the south from the union. This led to the proclamation of an independent republic based on the territory of the former PDRY. By then, however, Aden was under siege from the more powerful northern forces which were also making significant advances elsewhere in the south.

Thrown onto the defensive militarily and with diminishing prospects of defeating the northern advance, the southern leadership tried to mobilise the UN Security Council on its behalf. In the round of international diplomacy that followed, Saudi Arabia played a prominent role, successfully leading the effort to bring the case of Yemen before the UNSC at the end of May. This resulted in a unanimous call for a cease-fire and for negotiations between the two sides, much to the irritation of the northern leadership which insisted on the domestic nature of the conflict, fearing that UN involvement would mean *de facto* recognition of the secessionist regime in the south. In the event, only Saudi Arabia and some of its Gulf Cooperation Council (GCC) allies seemed willing to recognise the new state. This had more to do with Saudi hostility to the northern leadership, especially Salih, than with any great sympathy for al-Baidh in the south.

Indeed, such was the depth of Saudi hostility to the northern leadership that its efforts on behalf of the southerners reportedly went beyond diplomatic support to direct military assistance by shipping arms to southern ports. However significant this may have been it had little effect on the outcome of the fighting. By mid-June, after a short lull to allow the UN Special Envoy Lakhdar Brahimi to begin his mediation mission, fighting resumed in earnest. Thereafter, as talks between northerners and southerners proceeded sluggishly in Cairo, and as numerous cease-fires were brokered by various parties but broken by the forces on the ground, the siege of Aden intensified.

By early July, it was clear that the separatist state of the south was on the verge of collapse. With the port of Makallah and most of Aden in northern hands, the southern leadership's last-minute appeal to the GCC and to Egypt and Syria to recognise their rapidly disappearing state was a vain gesture. Realising the hopelessness of their situation, the surviving southern leaders fled Aden by boat on 7 July and the war came to an end.

Thereafter, it was left to the northern leadership to consolidate its conquest and to deal with the consequences of the war, both domestically and regionally. On the domestic front, after an initial period of looting, particularly in Aden, and the settling of scores with members of the old southern regime, prisoners of war were released and the administration operated more normally. To this end, Salih employed former officials of the PDRY who had fled to the north during the internecine fighting of the Yemen Socialist Party in 1986. Aden was declared the 'winter capital' of the state and it was clear that the northern leadership wished to avoid being seen as an occupying power in the south. However, both the northern and southern parts of the country faced a common economic crisis, compounded by the destruction each had inflicted on the other's infrastructure during the brief civil war.

Regional Implications

In the region, the position adopted by Saudi Arabia during the war further complicated the difficult relationship between Sana'a and Riyadh. Mutual suspicion and distrust already existed, based on such issues as the positions of the two countries during the 1990–91 Gulf crisis, Yemeni resentment of Saudi interference in religious and tribal matters, Saudi fears of Yemeni irredentism concerning the provinces of Asir and Najran and disputes over the ill-defined borders of the two countries. Against this background, Saudi assistance to and recognition of the secessionist regime in Aden could not easily be forgotten.

The effect was to complicate negotiations concerning the demarcation of the boundaries between the two states. The Taif Treaty of 1934, which had helped to define part of their common border, had not been renewed by the time it expired in 1994. By December, differences over the border issue led to an increasing number of border skirmishes between the armed forces of both sides, resulting in growing casualties.

Mutual accusations and recriminations soured the air between the two as Saudi Arabia accused the Yemenis of mounting armed incursions into Saudi territory, while Yemen accused Saudi Arabia of illegally encroaching on Yemeni territory. So serious had this become by the end of January that talks between the two sides in Riyadh, intended to resolve some of the issues, seemed to be on the verge of collapse and the Yemenis were accusing Saudi Arabia of massing forces on the border in an attempt to influence the outcome of the talks.

However, neither side could simply turn its back on the issue and both appeared to recognise the danger of the situation. The talks continued, and soon after they moved to Mecca in mid-February considerable progress was made. This resulted in the signing of a Memorandum of Understanding between the two states on 26 February which committed both countries to the Taif Treaty. The Memorandum also set up a number of joint commissions to define and demarcate the common border, as well as a joint military committee to prevent further armed clashes.

After a year of considerable upheaval and violence, it appeared that the government of the Republic of Yemen had succeeded in establishing its authority domestically and in reaching a *modus vivendi* with its powerful Saudi neighbour. In neither area can it be said that the present tranquillity necessarily means stability for the foreseeable future. There are too many imponderables and possible causes of conflict both domestically and regionally. Nevertheless, by March 1995 it appeared that the Yemeni government had found the resources to counter both potential sources of threat.

Asia

The almost total absence of substantive multilateral international relations in Asia has been a feature of the region for some time, but in 1994 the poverty of international affairs in this area of high economic growth was especially remarkable. The main reason for this trend was a major concentration on domestic affairs by most states, and especially the great powers. India and China were concerned about sustaining their economic reform, while China also suffered from the additional disability of an incipient jockeying for power as the death of its paramount leader, Deng Xiaoping, loomed. Japan continued to stumble towards political reform while suffering the effects of economic stagnation and destructive earthquakes.

To an important extent this focus on internal affairs was a way of not having to deal with some unsettling trends in the region's international affairs. While there was initial optimism that the Korean nuclear dispute had been defused, it soon became clear that the fundamental factors behind the dispute had not disappeared. Further south, China's seizure of new territory in the Spratly Islands reminded everyone in the region that they were living with a rising power with scores to settle. China's naval action, and continuing trade disputes with the United States, were vivid reminders that multilateralism, and indeed the future prospects for stability in the region, depended very much on whether China had the courage to move further and faster towards international interdependence in what was soon to be the post-Deng Xiaoping era.

Uncertain Weather Ahead in China

Just before Mao Zedong died in 1976, the much-quoted political aphorism about his anointed successor was 'with you in charge, I am at ease'. As China awaited the death of Deng Xiaoping, the turbulence of Chinese political life disrupted whatever ease might remain in the minds of those contemplating his succession.

The turbulence has obvious causes. China is buffeted by storms that emanate from the great heights to which its economy has soared. Its society and political system are pulled along at a dizzying speed. Not surprisingly, Chinese foreign policy has also come under strain. Of course these are all the risks and results of economic success, but they make for uncertain times nonetheless. Although it will probably not cause an explosion, or the break-up of China (as a report commissioned by the US Defense Department and released in February 1995 suggested), it is no exaggeration to say that China is undergoing a revolution.

Economic Shock Waves

In 1994 China once again led the world in economic growth. Such success is not generally considered a problem, but it was for a leadership that had promised it would achieve a 'soft landing' from 13.4% growth in 1993 (13.6% in 1992) to 8–9% in 1994. In the event, if the increasingly dubious government statistics were to be believed, the economy grew at 12% in 1994, suggesting that the government did not have a firm grip on the levers of economic control. This lack of control had many causes.

One obvious explanation for the inability to slow the runaway economy was the failure of reforms announced in autumn 1993 intended to control the money supply and tax system. Money was still being printed at excessive rates (hence inflation remained stuck at around 27% in early 1995, the highest ever under communist rule) in part because the government was unwilling to privatise loss-making state businesses. In fact, there were 1,000 more state-owned enterprises in 1994 than in 1993. Some 70% of all investment in China is in state enterprises, up from 61% in 1989, although the total contribution to national economic output of these enterprises has fallen (to 43% in 1994). The Chinese authorities feared that privatisation would lead to unemployment which could then lead to social disorder at a time of already great uncertainty about post-Deng China. Discretion and deception over statistics was considered the better part of economic valour, even though Chinese economists knew it was a risk they could take only for a short while.

Another reason for the loss of control of the money supply and tax system was that much of the explosive growth and tax evasion was taking place in the richer parts of coastal China. Local officials were unwilling to surrender power and money to a leadership in Beijing that few expected to survive the post-Deng era. For much of 1994, Vice-Premier Zhu Rongji, generally considered China's leading liberal, toured the country to muster support for his banking and tax reforms, but with seemingly little effect. The World Bank and a bevy of foreign business people with investments in China reported that Zhu's reforms had been blocked. Some argued that Zhu (once known as 'one chop Zhu' for his ability to implement such programmes in one go), could not be expected to succeed while the power struggle to follow Deng is under way. Others argued that the task was inherently difficult because economic power that had been decentralised for so long was unlikely to be surrendered to a central government that was patently incapable of taking the decision to cut the central deficit and privatise state enterprises.

Losing Control

Only the death of Deng could break the deadlock. While the country waited, authority and social control grew shaky. Central government admitted that at least 80 million peasants had recently moved from the countryside to the cities, and foreign analysts suggested that the figures were 50% higher. With such rapid social change in an equally rapidly modernising peasant economy, it was not surprising that social problems reached levels unimagined since the Communist Party seized control in China. According to official figures, the number of cases of serious crime surged by 15.6% in 1994 (Amnesty International reported that 2,500 Chinese were executed in 1993). Chinese officials also reported the seizure of hundreds of thousands of illegal firearms, and condemned the 'rampant' production of unlawful weapons. Chinese officials also became more open in reporting what many had noted already, that the Chinese environment was one of the casualties of the rapid economic growth.

In a society undergoing major changes, these were natural outcomes and not immediately threatening to continued economic growth. Of more immediate importance was the evident failure to control China's huge population. China's total population reached 1.2 billion in the second week of February 1995, five years ahead of China's optimistic planning. A decentralised economy seems to have allowed more people to flout the 'one-child family' policy.

Chinese leaders were uncertain how much to make of the risks of decentralisation. Although it was a convenient and accurate excuse for the ineffectiveness of government rule, the articulation of such an excuse also suggested that the government's authority was much weakened and therefore that it could be challenged or even ignored by domestic and foreign rivals. Unease was even evident over the record-breaking performance of Chinese athletes. In 1994, when world records were shattered at track and swimming events (at the Asian Games in October China won 137 gold medals), Chinese took great pride in their achievement. But by the end of the year, when it could no longer be doubted that drug-taking was behind the record-breaking, government officials switched from accusing the Western press of inventing stories out of envy, to blaming coaches and athletes who were pursuing their own 'training programmes'.

While cheating at sports might be dismissed as an insignificant detail rather than a widespread problem, it was far more difficult for central authorities to explain why even the armed forces were engaged in corrupt practices. It was well known that the People's Liberation Army (PLA) had for some time been involved in a number of enterprises not usually associated with military organisations, but most observers believed that it had escaped the debilitating effects of market reforms. By 1994, however, the steady trickle of stories about corruption in the armed forces and the widespread focus on expense accounts rather than expert military skills suggested that the PLA now had a major problem. Even foreigners were reporting that they had to pay bribes to PLA officials, and Hong Kong's fancy right-hand drive cars were appearing on the streets of China with the PLA at the wheel. Since the PLA was the party and state's chief law-and-order enforcer (it had done the dirty work on the streets of the capital in June 1989), PLA corruption highlighted a party and state in trouble.

Taken together, this revolutionary tinder scattered throughout Chinese society was clearly not just the creation of the Western media, although some tended to exaggerate the risks. Most of the data and the warnings about their importance came from Chinese academics and officials. For example, the elder statesman Yang Shangkun reported after an inspection tour in April 1994 that the state was failing to control 'lawbreaking and decadent behaviour'. Even the highest-level party and state meetings in 1994 provided ample evidence of the problem. At the seventh session of the standing committee of the eighth National People's Congress in May 1994, the authorities paid special attention to strengthening the public security apparatus.

In September 1994, the fourth plenum of the Party's 14th Central Committee focused on the problems of ensuring 'democratic centralism' in the age of reform – indicating that the provinces continued to flout central authority. All this was made more difficult by the uncertainty created by strong rumours that Deng Xiaoping was on a life-support machine. It was clear that the firm hand that had grasped the helm with such assurance was not only weakened, but had effectively been removed.

Caught in a Web of Interdependence

However much Chinese officials tried to reassure the outside world that the succession to Deng was already decided and stability would be assured, few foreigners were convinced. Chinese leaders, most notably President (and party leader) Jiang Zemin, were active on the international stage. Jiang visited Russia, Ukraine, France, most Association of Southeast Asian Nations (ASEAN) states and Vietnam. His appearance at the Asia-Pacific Economic Cooperation (APEC) summit in Bogor in November 1994 was touted as the peak of the year's diplomacy. Premier Li Peng visited South Korea, Germany and Romania, as well as touring Central Asia, Mongolia and Myanmar (Burma). Defence Minister Chi Haotian visited India. Zhu Rongji was also seen in Japan, parts of Africa and the Middle East, and most notably talking to leading executives at the World Economic Forum in Davos in January 1995.

The time when the outside world was impressed by the showy side of Chinese delegation diplomacy has long passed, however. Even the visit to China by Russian Prime Minister Viktor Chernomyrdin in May 1994 (only one of 40 heads of state or government to visit in 1994) passed virtually unnoticed. Far more important was what happened on the shop floor in China or in hard-headed international negotiation.

It is perhaps not surprising that during 1994 China had a hard time in international affairs. Data on foreign direct investment showed a sharp drop in actual funds spent. It also became clear that some earlier figures had been inflated by Chinese businesses taking advantage of the fact that even promises of investment attracted significant tax concessions and government support. Warning signs also flashed when it was revealed in May 1994 that more money was flowing into the United States from China than the other way around, following a trend set the previous year in the outflow of funds from China to Hong Kong.

The total stream of 'hot money' from China was estimated at $10–15bn each year, suggesting that Chinese economic growth may be uncomfortably similar to that in Latin America. Many money trails led to the families of senior leaders and members of the higher echelons of the PLA. Chinese officials were certainly anxious to refute suggestions of any similarities between conditions in their country and those in others such as Mexico during its financial crisis in early 1995.

The most acute international economic crisis for China was in the realm of trade rather than finance. China had confidently expected to join the new World Trade Organisation (WTO) as a founder member and left the matter of agreeing terms to the last moment. The American administration, with the support of the European Union (EU) and other Organisation for Economic Cooperation and Development (OECD) members, insisted that while China was welcome to join the WTO, it would have to do so under conditions that opened its markets and ensured its compliance with free-trade conditions and the new dispute-settlement mechanisms set up under the WTO. China argued that it was a developing country and should be treated leniently.

The Americans and the EU in particular were wary of repeating what they felt had been a similar mistake when Japan and the newly industrialised countries were allowed to join the international trading system, and felt they should insist that China joined in a way that ensured fair trade. China was sure it could appeal to developing countries and other Asians for support, but found that this only infuriated the Americans and the EU even more. In the end, China failed to agree terms and the WTO was established without it on 1 January 1995. Talks continue, and China could still be retroactively deemed a founder member.

Discussions on trade with China continue to be difficult. China does not seem to understand that conceding opening its economy to international standards and access is not a concession to the US, but a concession to common sense. China still sees the world in encrusted nationalist terms where nineteenth-century virtues of sovereignty are too sacred to be sacrificed. This difficulty lies at the heart of China's problem with bilateral trade, most visible in its row with the US over intellectual property.

After much heated rhetoric from both sides failed to achieve any solution, the US announced that it would impose Super 301 tariffs on $1bn-worth of Chinese exports to the US market unless China agreed to American terms in protecting intellectual property. China argued that it was unfair of the US to expect a developing country to control the lawless parts of its economy, and that the US was simply using this issue to force open the Chinese market. It is true that China ran a trade surplus with the US in excess of $22bn in 1994 (the precise figures are in dispute), but it is also true that China not only flagrantly violated copyrights for internal use, but most of its illegal enterprises were exporting to what would otherwise be US markets in Asia and even Europe and Canada.

The deterioration in Sino-US relations apparently came as a surprise to China. China believed that it had learned how to play capitalist against capitalist when it succeeded in beating off those who wanted to link access to the American market to China's human-rights record. It had forgotten that even under a Republican administration the US had always driven a hard bargain on trade issues. A post-Cold War America was prepared more than ever to implement a vigorous geo-economic strategy.

The deteriorating relationship between China and the US even extended to military issues. Despite the visit to China by US Secretary of Defense William Perry in October 1994, there was a notable shift in the US to more explicit concern about Chinese military power. The US Arms Control and Disarmament Agency publicised its estimate that Chinese defence spending was in excess of $100bn a year. The Pentagon, which had previously played war games in which China was the enemy behind firmly closed doors, made it known that it had held such an exercise near the end of 1994 (and that China had been the winner). More telling was the fact that Chinese ships and aircraft more than once deliberately challenged a US aircraft-carrier battle group in international waters. The most serious incident took place in the Yellow Sea near China on 27–29 October 1994. Chinese officials reportedly warned the Americans that they would 'shoot to kill' next time.

In this environment of growing tension, China's continuing refusal to join other members of the international community in pushing forward arms-control measures took on heightened significance. Beijing continued to test nuclear weapons in 1994, thus putting at risk the prospects for a Comprehensive Test Ban Treaty (CTBT) and, by association, the Nuclear Non-Proliferation Treaty (NPT), being reviewed in spring 1995. American officials are also unhappy about Chinese arms sales to Pakistan, enriched uranium sales to India in January 1995, and nuclear links with Iran.

In fact, to the US China was beginning to look far more like a superpower rival, at least in the sense that so many issues of international security required the careful handling of China. Nowhere was this more evident than in resolving the problem of North Korea's nuclear-weapons programme. American officials, who might have preferred to pressure North Korea through sanctions, were deeply unsure in autumn 1994 of the position China was prepared to take in enforcing a tougher sanctions policy against Pyongyang. Without active Chinese support there was little hope that North Korea could be persuaded to comply fully and immediately with International Atomic Energy Agency (IAEA) conditions.

A less pressing worry in 1994, but one that loomed larger when looking ahead to 1996, was Chinese attitudes to the increasing democratisation of Taiwan. In 1994, Taiwan's voters opted for a mixture of change and continuity. They backed the Democratic Progressive Party (which is committed to independence) in mayoral elections in Taipei, while voting for the ruling Kuomintang candidate for Governor and in other major elections. China reacted with an equally finely balanced message of pressure and patience when Jiang Zemin addressed the people of Taiwan just before the Chinese New Year in early 1995.

In summer 1994, Taiwan's government formally announced its intention to cease claiming the right to rule mainland China, thereby lending tacit support to those who saw Taiwan's future as one of creeping independence. Beijing continues to warn that any declaration of independence would be a *casus belli*, but is finding it difficult to stop the conditions for independence from developing. With the first popular presidential election looming in 1996, the task is set to become more difficult.

This is not the only fateful election scheduled for 1996. Not only will the presidential elections in the United States be vigorously contested by a resurgent, pro-Taiwan, Republican Party, but the last free elections in Hong Kong will be held. The terms of those elections were finally agreed by the Hong Kong Legislative Council (LEGCO) in July 1994, in the teeth of fierce Chinese opposition. Britain and China had held fruitless negotiations after Governor Chris Patten proposed widening the democratic franchise. China's immediate response to LEGCO's defiance was a promise to disband the democratically elected legislature and to impose its own system. China also began exploring ways to enlarge the authority of 'a second kitchen' which would in time become a rival, and even a more powerful, force in Hong Kong before 1 July 1997 when formal authority is to be handed over to China.

In early 1995 there were signs that China might be prepared to change the tone, if not the details, of its policy towards Hong Kong. Since LEGCO could no longer be induced to vote down the Governor's proposals, the level of vituperation eased, although policy remained designed to isolate the British executive. China even put forward, and vigorously supported, candidates of its own for the electoral contest in the colony in 1996 in the hope that it would win at the polls. The Chinese clearly recognised that electoral success would be a far more potent weapon in the long term in building support in, and control of, Hong Kong than authoritarian acts could possibly be. From Beijing's point of view, participation in the election could be cost free. If its chosen candidates suffered electoral defeat in 1996 China would blame the perfidious Patten and fulfil its promise to disband LEGCO.

If China's instinctive nationalism seemed a little more restrained over Hong Kong in early 1995, the same could hardly be said about China's pursuit of territorial claims in the South China Sea. Since 1974, when China seized islands in the Paracels from Vietnam, Chinese forces have been spreading their control further south as political and military conditions permit. Armed clashes in 1988 and on a smaller scale in 1992, were, like the 1974 events, at Vietnam's expense. While China and Vietnam continued to spar in the region (in April 1994 Vietnamese ships successfully chased off a Chinese one) the most dramatic action took place further south and east. In January 1995, some hapless fishermen from the Philippines were detained by Chinese forces when they stumbled across nine Chinese naval vessels, various new buildings and fluttering Chinese flags on the aptly named Mischief Reef, some 130 nautical miles off the coast of Palawan and more than 600 miles from the nearest Chinese territory.

China had struck again, this time picking on a Philippines that in the dying days of the Cold War had demanded the removal of US troops which had been stationed on its territory since 1898. Despite a 1992 accord with ASEAN states not to use force in the South China Sea, China calculated that neither ASEAN nor the US would stop it from taking the territory it claims. In this case China was right, and the uneasy shuffling by ASEAN and US officials suggested that if China exercises care it could continue to extend its reach in parts of the disputed Spratly Islands with only the mildest diplomatic resistance. The Philippines admitted that it would not (because it could not) eject the Chinese from the Reef.

Far from having to pay a price for its reef-grabbing, China found that foreigners were queuing up to sell it weapons to help it to extend its reach. It was revealed in late 1994 that Israel had helped China build the F-10 aircraft, a variation on the Israeli *Lavi* fighter. Israelis are now helping with new weapons systems for the aircraft, including *Rafael* missiles. Russians claimed to have sold China $2–3bn-worth of arms in 1994, and the first of the *Kilo*-class submarines arrived in February 1995. No Western arms were apparently yet on offer, but gradually improving military-to-military contacts with Western armed forces suggested that China might be able to learn more about what it needed to acquire from other sources.

Ambiguity Reigns

The arms sales to China, much like the passive response to the mischief on the reefs, was simply part of the natural uncertainty in the minds of outsiders about how to handle a rising and uncertain China. Chinese frequently and smugly ask foreigners if they want China to be strong or weak. The simple answer, of course, is that foreigners want a cooperative China, one that is tied into an interdependent world. How this difficult objective might be attained remains the major uncertainty.

Problems ranged from welcoming China as a major food and fuel importer, although causing prices to rise, to deterring Chinese nationalism, without isolating or making Beijing more dangerous. The Australian defence White Paper in December 1994 stated more explicitly what other Asians also recognised but were only willing to mutter under their breath: China was the rising power to watch and worry about. Americans, with their more robust trade policy and larger military power, were furthest out in front in trying to manage the problem of a strengthening, but potentially unstable China. It is not only the Chinese who will have to prepare for an uncertain future.

Japan: Old Wine in New Bottles?

Japan has been suffering from the after-effects of real and figurative earth-quakes. A major earthquake centred on Kobe in January 1995 killed over 5,000 people and brought major physical destruction and disruption to the western part of the Kansai region of Japan. Despite the bravery and forti-tude of the inhabitants, major questions were raised about the effect the reconstruction costs would have on Japan's nascent economic recovery and about apparent weaknesses in Japan's crisis-management systems. The uproar in the media about the reported failures of the disaster-relief organisations prompted some Japanese government officials to comment that the 'CNN factor' had for the first time had a major impact on Japanese politics.

Politically, Japan struggled to cope with the repercussions of the split in the Liberal Democratic Party (LDP) and its subsequent loss of office in 1993 after 38 years in power. In mid-1994 Japan went through three prime ministers in as many months, until one of the most bizarre twists in Japanese post-war political history brought together in one coalition the two parties that had been arch-enemies for the past four decades.

Japan's foreign partners remained confused and at times despairing. German Chancellor Helmut Kohl was only half-joking when he remarked on receiving newly installed Japanese Prime Minister Tsutomu Hata in May 1994 that he wished Japanese prime ministers stayed in office long enough for him to get to know them. Kohl's hopes were not to be realised. By the time he was to have attended the Group of Seven (G-7) economic summit meeting in Naples only two months later, Hata had been replaced by Tomiichi Murayama. Moreover, Murayama's international debut was not an auspicious one: it was marred by his absence from the first day of the summit deliberations, struck down by fatigue and nervous tension.

The unexpected emergence of Murayama, the first socialist prime minister of Japan since 1947, reflected the state of flux in Japanese politics associated with the disintegration of a political structure built up over more than three decades. Nevertheless, with the global economic and security order also in transition, Japan's inevitably introspective political and – particularly after the Kobe earthquake – economic preoccupations have not allowed it to play a defining role in global affairs.

Change and Change Again

In mid-1993, as the lengthy era of LDP rule came to an end, there had been some expectation that, despite being head of a disparate coalition government, new Prime Minister Morihiro Hosokawa would prove not only a breath of fresh air for Japanese politics, but also a catalyst for changing the Japanese socio-economic structure. But the Hosokawa coalition of small centrist and left-wing parties was united only in its dislike of the LDP. Once in power it proved an unstable marriage of convenience.

Hosokawa's efforts to evoke a Kennedy-style idealism for cleaning up Japanese politics and society came to nothing, as he himself was forced to step down in April 1994 because of his own connections with Sagawa Kyubin, the company at the centre of the major money-for-favours scandal which had earlier helped to undermine the LDP.

But even before his resignation, Hosokawa had seemed to be weighed down by the difficulties inherent in controlling and massaging a coalition which contained significant policy and personality differences. These intra-coalition tensions became more pronounced over the appointment of his successor. Hata, the former LDP faction leader who had precipitated the summer 1993 split within the LDP and who was now leader of the Japan Renewal Party, emerged as prime minister. But immediately after helping to vote him into office, the Social Democratic Party of Japan (SDPJ) walked out of the coalition in a dispute over tax reform, leaving Hata in an impossible minority government position.

A capable if colourless politician, Hata was given no opportunity to make any significant policy moves. Faced with a vote of no-confidence before he had had time to do little more than pass the overdue budget, Hata resigned at the end of June, thus ensuring himself a place in the record books as the shortest-ruling prime minister in Japanese post-war history. Behind-the-scenes manoeuvring reached fever pitch, with Ichiro Ozawa, Hata's right-hand man and the anti-LDP forces' chief strategist, doing his best to stitch together another anti-LDP coalition.

Ozawa overplayed his hand, as his clear dislike of the SDPJ helped to push it into the arms of its long-standing enemy, the LDP. The LDP, now headed by the reformist Ichiro Kono, was determined to return to power as soon as possible, but had to agree to support the socialist Murayama as prime minister in order to do so. A third party, the small LDP-splinter party *Sakigake*, was coaxed into the coalition to ensure a parliamentary majority. However, a number of LDP stalwarts balked at these moves. A few abstained in the vital Diet vote, while former Prime Minister Toshiki Kaifu, at the very last moment, actually left the LDP to be proposed as the anti-LDP parties' candidate.

The 70-year-old Murayama narrowly won the parliamentary vote against Kaifu, but he had to reward his new allies, the LDP, with significant cabinet positions to ensure his premiership. Kono became Deputy Prime Minister and Foreign Minister, and LDP politicians took most of the other key cabinet posts apart from that of Finance Minister, which was given to *Sakigake*'s leader, Masayoshi Takemura.

Papering Over the Cracks

Murayama, like his two immediate predecessors, has found coalition management a time-consuming and tiring process. He and his SDPJ colleagues had been forced to put aside considerable policy differences with the LDP, in areas such as defence, foreign, tax and economic policies, to set up the coalition in the first place. His willingness, though not without some personal anguish, to make a succession of conciliatory gestures, such as admitting that the Self-Defense Forces (SDF) were constitutional, kept the coalition together during the summer and autumn of 1994.

Murayama's policy reversals, however, unsettled many in his own party. Factional disputes within the SDPJ are not new, but their intensity is. By the end of the year, a group led by former SDPJ chairman Sadao Yamahana was on the verge of leaving the party. Only the Kobe earthquake made them hold back, but the SDPJ split has been merely postponed, not avoided.

At the same time differences amongst the opposition parties weakened their ability to exploit the policy differences between the LDP and the SDPJ or even within the SDPJ. Thus, to take advantage of the new electoral funding rules which favour larger parties and to avoid the LDP's own divide-and-rule tactics, in December 1994 a hotchpotch of nine opposition parties and groups formed one new party, the New Frontier Party (NFP or *Shinshinto*). With 178 members of the lower house, it became the second largest after the LDP (200 seats). But by choosing Kaifu as their new leader and Ozawa as their Secretary-General, the NFP looked like a collection of yesterday's men; their initial policy pronouncements tended to be bland and barely distinguishable from those of the LDP.

That the governing coalition began to look shakier in the early months of 1995 had less to do with the emergence of the NFP than with the intra-

SDPJ manoeuvrings and policy setbacks for the coalition. Frictions between the coalition politicians and the bureaucracy led to the unprecedented sacking of three Ministry of International Trade and Industry officials. Responses to the Kobe earthquake were slow and uncoordinated. Bank of Japan and Finance Ministry officials were reprimanded for overstepping the mark in trying to bail out two large but bankrupt credit unions. The government's proposed five-year programme of deregulation proved difficult to finalise. The Kobe earthquake not only exposed the inadequacy of government emergency services, but also threatened a new burden for the sluggish Japanese economy.

The economy, indeed, continued to be as great a problem for Murayama as it had been for his immediate predecessors. The gross national product (GNP) growth rate in 1994 staggered up to around 1%, better than 1993's negative growth record, but nowhere near the rates of the late 1980s. Income tax cuts, approved in autumn 1994 after prolonged political horse-trading, had little immediate impact on the economy. More importantly, a proposed five-year programme of deregulation proved difficult to finalise. A consensus has gradually been growing in Japan that loosening the grip of myriad bureaucratic regulations and red tape is the only way to restore Japan's economic competitiveness in the medium term, but business lobbies tend to temper their enthusiasm for deregulation when it directly affects their own protected sector. Politicians within the ruling coalition parties have their own vested interests too. Marked rises in the value of the yen against the dollar also slowed economic recovery, particularly by eroding earnings overseas; by March 1995 the yen had appreciated by 23% compared with two years earlier. The government seemed unable to do much to stem this problem.

All the government's economic plans for 1995, however, were wrecked by the Kobe earthquake. Estimates of the reconstruction costs varied widely, from US$20bn to $200bn. Most of the costs fell to the Japanese government, as the insurance companies were comparatively under-exposed. Companies in the Kansai area, including several of Japan's major export manufacturers, have found their production severely disrupted. But some sectors, notably construction and public works, are bound to receive a stimulus. It remains to be seen whether, in the medium term, the reconstruction of Kobe and the related attempts to improve earthquake-preparedness in other regions in fact act as the focus of a new national economic strategy.

Another Quiet Year Abroad

Neither the Hata nor the Murayama administrations had much time or inclination for new initiatives in foreign policy. Relations with Russia marked time. Japan was marginally more critical than in the past about

China's military build-up, but still agreed to a new yen loan for China. Japan did participate in the first meeting of the ASEAN Regional Forum (ARF) in July 1994, but was unable to win support for its proposal that nations should display greater transparency over military affairs by issuing defence White Papers. By agreeing to host the next annual meeting of APEC forum in November 1995, Japan took on the responsibility for putting some substance into the free-trade guidelines agreed at the November 1994 meeting in Indonesia. However, Japan's belief that economic cooperation should go hand in hand with trade liberalisation puts it at odds with several APEC members who see 'free trade' as the touchstone.

Relations with the United States, North Korean nuclear-weapon development and a permanent United Nations Security Council (UNSC) seat were the issues that dominated the year. Trade friction with the United States remained a lively topic, especially as Japan's bilateral trade surplus in 1994 was over $60bn for the second year running. The initial sympathy from President Bill Clinton and his administration for Hosokawa offering the best chance for effecting fundamental change in Japan had dissipated by the time they 'agreed to disagree' at a summit meeting in February 1994 over 'numerical targets' for key sectoral products. Hata, also, made no ground on the trade disputes, and it was left to Murayama to sort out the resulting mess. Complicated and painstaking sectoral negotiations on telecommunications, insurance, financial services and flat glass eventually brought agreement, but cars and car parts remained a sticking point. Murayama's summit meeting with Clinton in January 1995 was less acrimonious than Hosokawa's a year earlier, but, with a Republican-dominated US Congress keeping a close eye, yet more rounds of convoluted sectoral negotiations will be the order of the day.

The Americans were not able to draw much consolation from Japan's unresolved debate about its role in international politics either. There was a wide gulf within the Hosokawa coalition government between Ozawa's call for Japan to become a 'normal country' in global security affairs, and the SDPJ's criticism of even the existence of the SDF. Hosokawa leaned towards the pacifist end of the scale, but Hata was very close to Ozawa's thinking.

Murayama was forced into grudgingly accepting the constitutionality of the SDF, but made it clear that the Forces needed to be pruned in size and could not be used for dangerous UN peacekeeping operations (he quietly shelved the report of an *ad hoc* committee originally set up by Hosokawa which advocated full participation in UN peacekeeping forces). This meant that the rise in the 1995 defence budget has been pared down to 0.85%, its lowest level for more than three decades, and that the SDF was not sent to war-torn former Yugoslavia, despite the urging of Yasushi Akashi, the UN special representative there. Furthermore, Murayama played down the campaign for a permanent seat on the UN Security Council which had been launched by Hosokawa and pushed particularly forcefully by Hata. The Foreign Ministry remained enthusiastic about the bid, but Murayama's caution reflected not only his personal doubts, but also a general wariness amongst UN member-states to put their full weight behind the Japanese campaign.

The debate about the SDF's role was, however, thrown into new relief by the Kobe earthquake. Strict regulations governing the deployment of the SDF within Japan prevented the arrival of troops for emergency rescue activities until seven hours after the quake had hit. Criticism of the Murayama administration's slow response to the earthquake has provoked a review of Japan's overall information-gathering and crisis-management systems. The net result should be an improvement in Japan's foreign policy-making, as well as its domestic policy-making, capabilities.

But the most crucial foreign-policy issue which, indeed, all three coalition governments have faced has been the situation on the Korean peninsula, which has been particularly uncertain since the death of Kim Il Sung. As the Japanese 1994 Defense White Paper laid out, the old Soviet threat has been replaced by North Korea, potentially armed with nuclear weapons and, at the very least, possessing missiles capable of reaching Japanese territory. Indeed, this new realisation of vulnerability helped tip the balance in favour of participation in the US-proposed Theater Missile Defense project to develop an anti-ballistic-missile defence system.

The Hosokawa and Hata cabinets clearly hoped that the whole problem would go away, but kept close to the South Korean–US position on international inspections of the North's nuclear facilities. Murayama, whose own party has long had close contacts with the North, made known his preference for negotiations over sanctions and was relieved when the US–North Korean 'deal' was reached in October 1994. Japan seemed likely to follow the US steps towards quasi-recognition of North Korea by re-starting its own diplomatic negotiations with the North which had been suspended in 1992, but wrangling amongst coalition partners delayed an early initiative.

As the North Koreans, amongst others, were not slow to point out, Japan has been in some difficulty over its own nuclear policies. In June 1994, Prime Minister Hata admitted publicly and for the first time what nuclear experts around the world have long suspected, that Japan did have the technological capacity to make nuclear weapons. American officials quickly warned the Japanese that should it become evident that they were indeed moving towards acquiring nuclear weapons, the US would cut down its technological assistance to Japan on its fast-breeder reactor programme. Japanese officials hastened to reaffirm that Japan had no intentions of making nuclear weapons. Hata's successor, Murayama, while quietly playing down his own party's long-held opposition to even the peaceful use of nuclear power, did try harder to reassure neighbouring countries through greater openness; for the first time Japan revealed accurate details of its plutonium stocks, held both inside Japan and in facilities in the UK and France.

But controversy about Japan's military activities – past, present and future – is set to be an enduring theme of 1995, the fiftieth anniversary of the end of the Pacific War. Although Murayama has followed the precedent set by Hosokawa in personally apologising for the past, two cabinet ministers were forced to resign in 1994 for comments seen as whitewashing past militaristic activities, compensation claims by the 'comfort women' remain unsettled, and a group of over 150 LDP politicians is lobbying against any formal parliamentary apology for the war. The announcements of planned exhibitions, memorial services and even postage stamps in the United States – especially those recalling the nuclear attacks on Hiroshima and Nagasaki – have already raised tensions with Japan. As Asian neighbours also move to celebrate their liberation from Japan, the Japanese government may well find that its planned $1bn fund for cultural and youth exchanges with its Asian neighbours will not be enough to assuage the past.

Stalemate on the Korean Peninsula

This was a long-anticipated year, the year when Kim Il Sung, the only leader North Korea has known since 1948, finally passed from the scene. And nothing happened. Indeed, so little happened that some observers became worried at the lack of activity. Yet the transition process by which Kim's son, Kim Jong Il, took over his father's role appeared to work smoothly. Earlier scenarios in which the death of the father led to an open succession struggle, widespread chaos, then swift collapse have thus far proved false.

It was also a year of continued nuclear tension. The on–off US–North Korean negotiations appeared to have broken down irrevocably in spring 1994. Then suddenly they were back on track following the intervention of former US President Jimmy Carter. No sooner were talks under way again, than Kim Il Sung's death seemed destined to lead to their indefinite postponement. In fact, they resumed with only the briefest of pauses and an agreement was signed in October 1994. There were, however, many unresolved details, and the issue was by no means settled.

All other developments in both North and South faded away against this background. South Korean reports continued to claim that the North's economy was still sinking, but there were stray signs at the end of 1994 that perhaps the worst was over. In the South, some of the steam went out of President Kim Young Sam's reform programme. A series of accidents also cast a shadow. But the economy picked up dramatically despite a drought which affected agricultural output.

North–South relations, however, went backwards. During the Carter visit Kim Il Sung proposed that there should be a summit between him and Kim Young Sam. Talks to achieve this had begun when the 'Great Leader' died. When the South declined to send condolences, the North reacted with angry denunciations of Kim Young Sam and refused to continue the discussions. Thereafter, the North pursued its links with the US, ignoring the South as far as possible.

North Korea: the Great Leader Goes

It was bound to happen some day, but when Kim Il Sung died of a heart attack on 8 July 1994, the world could not be sure what would happen. Many had predicted that Kim's death would mean the immediate collapse of the regime or some other catastrophe. Few thought that Kim's designated successor, his reclusive son Kim Jong Il, would be able to manage the transition. There were lurid rumours about the diseases and illnesses from which he suffered.

In fact the transition appeared to take place without a hitch. North Korea went into paroxysms of mourning, which even nature joined according to the North Korean media. Kim Jong Il made few appearances, thus increasing speculation about his difficulties in assuming office; there was also a renewed spate of stories about illnesses ranging from brain damage to diabetes. Yet the machinery of government seemed to function. US negotiators on the nuclear issue found that exchanges carried on as before, with the same team of officials and with only slight delay. Kim Jong Il's absences from view, which caused speculation in the world's press, were explained away by North Korean officials as the respectful mourning of a dutiful son. In the absence of any better reason, and given the high esteem traditional Korean society attached to filial duty, this had to suffice.

North Korea thus entered the post-Kim Il Sung era with neither a formal head of state nor a leader of the Korean Workers' (Communist) Party, a condition still in existence in March 1995. Among the oddities that make up North Korea, these omissions are perhaps less odd to a domestic audience than to outsiders. Some outside observers felt that this behaviour strongly implied that a struggle for power was under way behind the scenes. There were other signs, however, that suggested that the younger Kim was the inheritor of power. There were frequent references to 'Kim Il Sung is Kim Jong Il and Kim Jong Il is Kim Il Sung', while the younger Kim's birthday in February became the main national holiday, replacing 15 April, Kim Il Sung's birthday.

Rumours of family or other opposition continued, but the evidence was still lacking. Another group sometimes portrayed as a likely focus of opposition is the military. That they are important was shown by the fact that most of Kim Jong Il's relatively few appearances after the death of his father were linked to the military, but there were few other clues to their role. The death of the veteran military leader O Jin U in February 1995 either removed a strong supporter, according to some, or a strong opponent, according to others. Whichever it was, Kim Jong Il attended his funeral. It is too early to rule Kim Jong Il either fully in, or fully out of, control in North Korea. He seems to have managed a short-term grasp on power, but unless he fills the senior positions left vacant by his father's death, his long-term prospects will decline considerably.

'It's the Economy, Stupid'

Economically, the North continued to flounder. There have been no signs from the leadership thus far of a fundamental change of approach to economic matters, which might suggest that the regime is unable to agree on such changes. It could equally mean that if changes are contemplated they will be put into effect gradually and circumspectly. Various clues point in that direction. There has been some reorganisation of economic ministries. North Korean companies abroad appear freer to make deals where they can. There has been continued publicity for the Rajin–Songbong development zone, though few signs of actual investment yet.

Yet the basic problems remain. The economy is still heavily centralised. Military expenditure continues to account for about 25% of the gross domestic product (GDP). Statistics from South Korea pointed to a further decline of about 4.3% in the North's economy, the same as in 1993. This would represent the seventh consecutive year of decline. Anecdotal evidence told of darkened streets, roads empty of traffic and factories barely functioning. There were also continued reports of food shortages. The North Koreans reported a good harvest, but since harvests were poor in most of the rest of Asia, this was greeted with some scepticism. North Korea continues to export foodstuffs to China, Russia and, indeed, to South Korea; in the latter case they include fish and potatoes, and shipments of North Korean apples are planned.

There were other signs of an upturn in North–South trade towards the end of 1994, despite unrelenting North Korean propaganda attacks on the South's leaders. The little-publicised trade in textiles and other finished goods, as well as foodstuffs, continued to grow. As a goodwill gesture following the October 1994 Geneva nuclear framework agreement, South Korea lifted some of its restrictions on trading links with the North in November. The North appeared to reject this, with much vehemence. Yet there were immediate indications of great South Korean interest in the

North's markets. Several of the South's big conglomerates, the *chaebol*, quickly organised visits to the North.

The Ssangyong group led the way, but others were not far behind. None seemed to want to confine their operations to the small-scale consumer-related enterprises envisaged in the South's National Unification Board guidelines. They preferred telecommunications and electronic goods manufacturing. Smaller South Korean companies may pick up the slack. This eagerness to move into the North may eventually cause concern in the South, but for the moment, the more open policy towards North–South trade continues. South Korean sources also noted a slight increase in foreign trade generally at the end of 1994, having reported poor returns for most of the year.

There were also signs of a more open attitude to the outside world beyond the economic sphere. Several US delegations visited North Korea; one included James Lilly, former US ambassador to both Seoul and Beijing. Attempts were made to begin a dialogue with a number of countries, though the nuclear issue tended to overshadow these attempts. The Chinese were assiduous in balancing their increasing links with South Korea with delegations and exchanges with the North, although there seemed to be little substance in the contacts. Even the Russians began to talk about increased trade.

Relations with Japan hardly budged. Having brought the US to the negotiating table, the North Koreans seemed less interested in a better relationship with Japan. They may well assume that if they can move the US, the 'client states', Japan and South Korea, will have to follow. Like other neighbouring countries, the North Koreans were critical of Japan's shortcomings in handling the 'comfort women' issue and its alleged refusal to accept proper responsibility for its behaviour during the Pacific war. Japan's interest in a permanent seat on the UN Security Council and other international ambitions came in for attack, as did its alleged nuclear ambitions.

South Korea: the Elected Leader Fades

It was not an easy year for Kim Young Sam. Some of the euphoria of a civilian government began to wear off. This was inevitable, perhaps, as the first phase of reforms passed into history and it became harder to find new targets. The North Korean question, and especially the issue of how much say the South Korean government had over matters of vital concern to it which were being settled by US negotiators, was a constant sore. A series of accidents, none of which could really be blamed on the current government, beginning with the collapse of one of Seoul's main bridges with heavy loss of life, were seen by many as evidence of malaise at the centre. Switches in cabinet personnel, and a wholesale re-shuffle in December, were seen as desperate attempts to refurbish the President's image. Problems within the ruling party did not help.

Kim Young Sam's image as a reformer began to flag as the easy gains of the first few years proved hard to sustain. Obvious targets, such as the politicised military, corrupt politicians and corrupt money transactions, are out of the way, and it is not easy to find new targets without alienating support. There is no evidence that Kim was prepared to launch a wholesale attack on the structures of political power in the Republic of Korea (ROK), while areas such as reform of the state security apparatus and the National Security Law were effectively ruled out because of North Korean demands that they be ended. There were persistent rumours that Kim Young Sam had no more policies to put forward and was therefore relying too much on the bureaucracy. There were also signs that faced with a muck-raking press and apparently pro-North Korean students, the government's instinct was less liberal than it had first seemed.

Many in South Korea were particularly dissatisfied at what was seen as its marginalisation by the United States in discussions over the North Korean nuclear issue. Feeling put upon by great powers has long been a Korean grievance, but the high-profile negotiations of 1994 have brought it to the fore again. The US tried to sugar the pill with frequent consultations but inevitably, after the US–North Korean Agreed Framework was signed in October, South Koreans felt that they were being asked to foot the bill for an agreement to which they had been able to make little or no real input. Equally inevitably, some of this grumbling washed up at the President's door.

Other things were also going wrong. Corruption had not been eradicated after all; as the year progressed, more and more cases emerged of massive tax and other frauds, leading to the resignation and replacement of a number of mayors, including those of Inchon and Seoul. The government lost two by-elections in August in previously safe seats. Investigations of the 1979 military takeover which brought President Chun to power were abandoned, leading to an opposition walk-out from the National Assembly. This did not last long, although the ruling Democratic Liberal Party (DLP) took the opportunity to pass the budget unopposed, awakening memories of past government–opposition stand-offs. There were further echoes of the past in March 1995 when the opposition barricaded the National Assembly Speaker in his home locking him out of the Assembly in order to prevent debate and a vote on the ruling party's proposals on the local election law. Even though agreement on the election law was finally reached two weeks later, the episode did not help Kim Young Sam's image.

South Koreans detected a wider malaise in society. There were murders and robberies, some involving the military, which led to soul-searching about the way South Korean society has developed in recent years. Although the fault, if any, for the series of accidents lay in the past, the government was held responsible and its 'moral authority' was damaged by these events. Finally, there were troubles within the ruling DLP leading to the departure of Kim Jong Pil, whose political involvement stretched back to the 1961 coup. Kim Jong Pil announced that he would start a new party whose aim would be to introduce cabinet government. He was replaced in the DLP by a former general, Lee Choon Koo.

Kim Young Sam responded to these problems in much the same time-honoured fashion – he frequently shuffled his cabinet team. The pattern was set when the Agriculture Minister was sacked in April, a scapegoat for the ROK's problems with the Uruguay Round of the General Agreement on Tariffs and Trade (GATT). Later the same month the Prime Minister went. Others followed, with a clean sweep of the foreign-policy team in December, partly out of frustration over the US–North Korean issue. By that time, only one member of the Kim Young Sam original group of new-blood academics remained. Kim Deok had moved from being Director of the Agency of National Security Planning (NSP, formerly the KCIA) to be Deputy Prime Minister and Minister of Unification. Unfortunately, only days after his new appointment, he too resigned following revelations that the NSP had continued to monitor domestic political developments. By early 1995, the new blood of 1993 was largely on the carpet, and key areas such as foreign policy and relations with the North were safely back in the hands of the bureaucrats. These were perfectly capable hands, but they did not represent change and reform in the way that the first appointments had.

As he has done ever since he took office, Kim travelled. He went to Japan and China in March 1994, and to Russia and Uzbekistan in June. He attended the APEC summit in November and then went on to Australia, returning with a new vision of the ROK embracing 'globalisation'. His new cabinet was charged with introducing this concept, a task only hampered by the President's failure to define what he meant. Before long, all policies were being ascribed to globalisation, but it remained an elusive concept. Visitors to the ROK included US Defense Secretary William Perry in April, when the crisis over North Korea's nuclear programme seemed at its height. Japanese Prime Minister Murayama came in July, and Chinese Premier Li Peng at the end of October. Li Peng's main concern was to attract ROK investment in China; the South Koreans were disappointed that he ducked political issues whenever possible.

The good news was the economy. After several dull years, the ROK took off again. Growth was the order of the day, with manufacturing output up by 10%, consumer spending by 7.5% and exports by 17%. GDP as a whole was up 8% and inflation was below 6%. There were more moves towards trade liberalisation and more were promised. The down-side of the manufacturing boom was its attendant imports boom, as capital equipment flowed in to feed the growth, with a marked increase in ROK imports from the US, Japan and, a poor third, the EU. Wages rose on average by 15%; this economic recovery is apt to mean even higher wage demands in 1995. More consumer spending, and tales of murder and robbery, led to the moral outlook of the country being questioned, but most people seemed to welcome being richer. As a symbol of this new wealth, the ROK finally signed off from the World Bank's International Development Association.

Halfway through his term, Kim Young Sam faces an uncertain future. His personal popularity is still high, but that of his administration has fallen drastically. 'Globalisation' remains a nebulous concept with which to combat the pressures from North Korea, but the realities of finance may give the ROK more leverage in its international relations in the coming year.

The Nuclear Issue and North–South Relations

Most of 1994 was overshadowed by the ongoing nuclear crisis, and North–South relations stood still. The year began with a continued stand-off between the IAEA and North Korea over the implementation of the safe-guards agreement which the North Koreans had reluctantly signed in 1992, but which they had signally failed to implement. The United States spent most of the year trying to restart the talks which had stalled in 1993. Matters came to a crisis in May when the North Koreans announced their intention to remove the fuel rods from an experimental nuclear reactor. There were strong IAEA protests since this would effectively prevent the acquisition of information that would allow the Agency to deduce the history of the reactor's previous use. The North Koreans, ignoring the IAEA, went ahead anyway. The issue was brought by the IAEA to the UN Security Council where the United States pressed for sanctions to be mounted against the recalcitrant regime. The North responded that sanctions would be treated as an act of war.

Although North Korea may have been bluffing, the US decided to increase its own military position on the Korean peninsula. A long stale-mate seemed inevitable, though it soon became obvious that there was little regional enthusiasm for sanctions and their possible consequences. The crisis deepened on 14 June when North Korea informed the UNSC that it was withdrawing from the IAEA. At this point, former US President Jimmy Carter emerged as the *deus ex machina*. He crossed into North Korea from the South on 15 June, met Kim Il Sung and, after a day's talks, announced that North Korea was willing to freeze its nuclear programme in return for economic and diplomatic links with the US. The idea of such a package was not entirely new, but this time the North Koreans seemed far more serious about it than ever before.

The US administration was surprised and in some circles concerned about the Carter intervention, but it provided a way forward. Talks began, and even Kim Il Sung's death occasioned only a momentary pause. By October, an agreed package was ready. The North Koreans were to receive two light-water reactors (LWRs), assistance with other energy supplies and normalisation of relations with the US. In return, they would rejoin the IAEA, freeze all work on their current nuclear programme, begin disman-tling their existing reactors and, after several years, open all their nuclear facilities to inspection. An international body, the Korean Energy Develop-ment Organisation (KEDO), would arrange finance and oversee the intro-duction of the new technology into North Korea.

The US administration argued that the North was now following a US agenda and would only benefit if it stuck by the agreement. Others were more sceptical, seeing the agreement as a reward for the North's intransi-gence. The South Koreans were unhappy, particularly as the US was now in direct contact with the North and had pledged to improve relations. They became even more unhappy as they began to suspect that they would be the main contributors to KEDO, but drew consolation from the hope that they would supply the LWRs to North Korea.

Implementation of the agreement began well. The North suspended work at existing sites, and IAEA inspectors were allowed to resume moni-toring. But before long another crisis seemed imminent, as the North refused to accept that the LWRs would come from South Korea. The US insisted that there had never been any question of other reactors since only the South Koreans were willing to provide the necessary funds. By mid-March 1995, the North was threatening to end the agreement, the US administration was claiming that it had made clear all along to the North Koreans that South Korean reactors had been the only option, and the world braced itself for another round of brinkmanship. Among the new complications in the game is a Republican-dominated US Congress, which reluctantly accepted the original deal, but is unlikely to play a positive role in helping the Clinton administration to resolve its latest stand-off with the North Koreans.

It was not an auspicious background against which to play out North–South relations. As concern grew in the early part of 1994 over the nuclear stand-off, tension mounted in the ROK. The North reacted violently, claiming that sanctions would be tantamount to war and would reduce Seoul to a 'sea of flames'. Rhetoric this may have been, but with Seoul well within North Korean missile and aircraft range it had an uncomfortable ring. The ROK wavered between denying the existence of a crisis – 1994 had been billed as 'visit Korea year' – and staging full-scale anti-nuclear

contamination exercises. It also publicised the arrival of US military rein-
forcements in South Korea. There was some surprise when these activities
led Koreans into a wave of panic-buying and hoarding.

Carter's visit to Pyongyang defused the crisis. The South Koreans
were not happy either at the re-emergence of former President Carter on
Korean matters – for many still resent his plans for troop reductions in
Korea in the 1970s – or at his solution to the nuclear issue, but they
welcomed the offer of a North–South summit which he brought out of
Pyongyang. This was planned for 25–27 July. Talks on the practicalities
were under way when Kim Il Sung died.

The North postponed the talks. The South expressed the hope that a
summit would take place but, despite calls from some opposition MPs,
declined to express condolences to the North on Kim's death. Perhaps it
was politically impossible to express regret at the death of one who had
caused so much suffering. Yet Kim Young Sam had been willing to meet
the deceased Great Leader and shake his hand in life, as many senior
South Koreans had done since the first contacts in the 1970s. To some,
therefore, the government had missed an opportunity.

There was a further contradiction: South Korean commentators, who
had hitherto dismissed Kim Jong Il as mad, bad and dangerous to know,
suddenly discovered that perhaps it might be possible to do business with
him. However, instead of a gesture of conciliation, the government began
a witch hunt of students who, it claimed, were members of a pro-North
movement organising memorial altars for Kim Il Sung. Professing outrage
at the South's actions, the North cut off contact; it was not until February
1995 that the North–South telephone link was restored. Media denunciations of the South's leaders and Kim Young Sam in particular increased in
a crescendo.

The North had probably decided that it was doing so well in building
up its contacts with the US that it could ignore the South. It continued to
undermine the 1953 Armistice arrangement, which it wishes to replace
with a peace treaty with the United States. In May, North Korea announced it was withdrawing from the Military Armistice Commission
(MAC). It has argued that the MAC's work had been paralysed since the
appointment of a South Korean general as the UN commander in 1992. In
practice, while North Korean troops in the Joint Security Area at
Panmunjon no longer wore MAC insignia, they continued to deal with the
UN. In September, the Chinese, perhaps under pressure from North Korea, announced that they too would leave the MAC. In February 1995, the
North Koreans forced the Polish delegation to the Neutral Nations Super-
visory Commission, a part of the MAC machinery, to withdraw.

The South could do little but express concern at these moves, but took
comfort in the fact that the US – the main non-South Korean element in the
'UN forces' – insisted that the MAC was still in existence and refused to

deal directly with the North except through MAC machinery. However, even that reassurance was undermined in December. In an effort to recover the crew of a US helicopter shot down over North Korea, the US took advantage of a visit by Congressman William Richardson to Pyongyang, and other direct channels to secure the release of one crew member and the return of the other's body. While the US had used the MAC machinery as well, the South expressed concern at this willingness to move outside the Armistice setup.

The Outlook

After years of relative, if precarious, stability the future prospects for the Korean peninsula in 1995 are full of new uncertainty. Much depends on the shape of the new leadership in the North, but that remains as opaque as ever. There are signs that the South Korean leaders may adopt a tougher stance towards the North to regain some of the ground they feel they have lost through direct US–North Korean contact. The US administration, faced with a hostile Congress, may find it harder to pursue a conciliatory line in the face of renewed North Korean nuclear brinkmanship. The hope is that the North's new leaders will accept that they now have the best bargain they are likely to get, but the omens are not good.

Multilateral Security in Asia-Pacific

Two initiatives adopted in 1993 were each taken one small step further in 1994. The initiatives were evidence of a general willingness to explore the merits of multilateral security dialogues, but within strictly limited terms. They indicated also the extent to which, since the end of the Cold War, South-east Asia has lost its utility as a strategic concept. ASEAN, with American and Japanese support, has merged the region into a wider Asia-Pacific frame of reference. One purpose of the merger has been to cope with a new strategic environment in which the United States no longer commands confidence as a protecting power and China looms larger as a threat. In this new environment, the ASEAN states have shown themselves prepared to contribute to the regional balance of power only indirectly.

The ASEAN Regional Forum

The ARF held its first working session in Bangkok in July 1994 immediately after the annual meeting of ASEAN foreign ministers. The working session among 18 foreign ministers, including a representative from the EU, lasted three hours which permitted only brief declamatory statements and minimal discussion. The meeting was described as historic by its chairman, Thailand's Foreign Minister Prasong Soonsiri, who also claimed that it signified the opening of a new chapter of peace, stability and cooperation. Although the Cambodian civil war and the Spratly Islands dispute were touched on, the only regional security issue referred to specifically in the chairman's statement at the end of the meeting was that of the Korean peninsula in the context of the non-proliferation of nuclear weapons. The member governments welcomed the continuation of US–North Korean negotiations and endorsed the early resumption of inter-Korean dialogue. The South Koreans were party to the deliberations and the chairman's statement, but the North Koreans were not.

The underlying point of the diplomatic exercise was to make member governments comfortable and content with the ARF as an evolving framework for managing regional tensions. It was agreed that 'as a high-level consultative forum, the ARF had enabled the countries in the Asia-Pacific region to foster the habit of constructive dialogue and consultation on political and security issues of common interest and concern'. That collective judgement was certainly exaggerated given, for example, China's insistence that it was only prepared to discuss contending claims for jurisdiction in the South China Sea on a bilateral basis. Nonetheless, the general tone of the meeting was positive with no government striking a discordant note. In addition, it was agreed to convene the ARF on an annual basis and to instruct senior officials to collate and study papers presented and ideas raised on confidence- and security-building in Bangkok for the next meeting of foreign ministers in Bandar Seri Begawan, Brunei, in July 1995. The only specific proposal made to that end was to encourage all ARF countries to participate in the UN Conventional Arms Register.

With the emergence of separate ARF machinery in the form of a senior officials' meeting convened towards the end of 1994 in Brunei, ASEAN risked becoming subordinate to the institution it had helped to create. ASEAN's own enlargement was made possible in July 1994 by the agreement to admit Vietnam to full membership 12 months later. In January 1995 Cambodia sought permission to adhere to the Association's Treaty of Amity and Cooperation which carries with it observer status and candidate membership.

Asia-Pacific Economic Cooperation

The second meeting of heads of government and economics ministers of the 18-member APEC took place in Bogor, Indonesia, in November 1994 with an agenda that focused on freer regional trade and investment. A two-step approach was agreed whereby industrialised economies would achieve the goal of free and open trade and investment no later than 2010 and developing economies no later than 2020. The goals were not underpinned with specific measures, a task left to the next summit in Osaka in 1995. Much of the impetus for an accord on general principles came from the host government of President Suharto, which saw the occasion as an opportunity to demonstrate the international standing of Indonesia while it chaired the non-aligned movement. The occasion contributed to the general climate of regional security and during the meeting Chile was also added to APEC's membership. No attempt was made to duplicate the agenda of the ASEAN Regional Forum or to establish any lines of communication with it. But APEC shares with the ARF an underlying assumption about the interrelationship between economic cooperation and regional security. Once again, the only obvious discordant note was struck by Dr Mahathir Mohamed, Prime Minister of Malaysia, who issued a set of reservations after the Bogor meeting to the effect that his country would only commit itself to undertaking further liberalisation on a unilateral basis 'at a pace and capacity commensurate with our level of development'. He also set conditions for his attendance at the meeting in Osaka in 1995.

Problems of Regional Cooperation and Order

ASEAN governments were disappointed when US Secretary of State Warren Christopher was unable to attend the ARF in Bangkok in July 1994 because of the higher priority of the Middle East peace process. President Clinton did attend the APEC summit in Bogor in November amid regional concern about the security commitment of the United States after the reverses suffered by the Democratic Party in the Congressional mid-term elections. Ironically, just prior to the APEC summit, the Bangkok government had rejected an American plan for pre-positioning six military supply ships in the Gulf of Thailand for use in possible regional crises, including the Middle East and Korea. Thailand, with Malaysian and Indonesian support, opposed the Equipment Aboard Ships programme on the grounds that it would 'generate regional tensions', interpreted as concern over the likely reaction of China. Soon after the APEC meeting, which followed a visit by President Clinton to Manila, the government of the Philippines changed its mind about signing an acquisitions and cross-servicing agreement with the United States which was already in draft form. The rejection of a fairly routine agreement to facilitate port calls, in part because of its provision for storage services, reflected continuing Filipino sensitivities over harbouring an American military presence which had been withdrawn completely in November 1992.

Despite such rejection, the US has felt the need to offer assurances about its regional military commitments. Admiral Richard Macke, Commander in Chief of the US Pacific Command, did so in a newspaper article in January 1995. Later that month Thomas Hubbard, Deputy Assistant Secretary of State for East Asian and Pacific Affairs, visited Brunei, Malaysia and Singapore to secure funding for the US–North Korea framework agreement to replace Pyongyang's plutonium-producing nuclear reactors and to re-state the American commitment to regional security. In early March 1995, Joseph Nye, Under-Secretary of Defense for Planning, reaffirmed that the US intended to retain its 100,000-strong presence in Asia-Pacific.

ASEAN governments have been careful to avoid provoking China and so prejudice its continuing participation in multilateral security dialogues under the aegis of the ARF. The Chinese government encouraged this view, for example, when President Jiang Zemin visited Singapore in November 1994. Dr Mahathir Mohamad had been most outspoken in counselling against provoking China. Just prior to the APEC meeting, he interpreted the American proposal for pre-positioning military supplies as equivalent to establishing a military base. He pointed out that if countries in South-east Asia were to form a pact and single out certain countries as potential foes, then those countries would become their enemies. In a well-publicised speech in January 1995, Dr Mahathir argued that 'to perceive China as a threat and to fashion our security policy around this premise would not only be wrong policy, it would also be a bad and dangerous one'.

The regional state with most reason to fear China's intentions has been Vietnam, which is in contention over jurisdiction in the South China Sea. Vietnam has sought external countervailing support, and an agreement was reached between Vietnam and the United States in May 1994 to establish liaison offices in Washington and Hanoi. The pact was implemented in early February 1995, but with the American reservation that this would not constitute the establishment of diplomatic relations between the two countries. In June 1994, Vietnam and Russia concluded a friendship treaty which replaced that signed with the Soviet Union in 1978. Permission was obtained for the Russian Navy to continue to use Cam Ranh Bay, but its vestigial presence is not of any strategic significance. The intriguing prospect of an American military return to Cam Ranh Bay was raised in October 1994, when Admiral Macke visited Hanoi and said: 'I'm a naval officer and naval officers are always looking for good ports'. Tomiichi Murayama paid the first official visit to Vietnam by a Japanese prime minister in August 1994.

Relations between Vietnam and China have remained chilly as neither has given any sign of willingness to compromise over their competing claims for jurisdiction in the South China Sea. Tension revived in April 1994 when the Crestone Energy Corporation of Denver announced that it had begun exploring for oil within the area of an offshore concession granted by China in May 1992. Vietnam maintains that the area falls within its continental shelf. Concurrently, Vietnam signed contracts with a consortium of American and Japanese oil companies led by the Mobil Corporation to explore an adjoining westerly field within the area of China's maritime claim.

An unsuccessful fourth round of talks on the South China Sea in mid-August was followed later in the month by reports that both the Chinese and Vietnamese navies had harassed each others' supply and research vessels. Tensions moderated somewhat in November when President Jiang Zemin paid a state visit to Vietnam. In the concluding joint communiqué the two countries agreed to 'refrain from all acts that make things more complicated or broaden conflicts'. Plans were announced for a new expert group to address the rival maritime claims, and regional neighbours were assured that China and Vietnam were both willing to work towards a peaceful settlement in a significant gesture of support for the ARF process.

Apprehension over the intentions and good faith of the People's Republic revived in early February 1995, however. The government of the Philippines claimed, backed by photographic evidence, that China had erected four structures on Mischief Reef , some 200km west of the island of Palawan and within the area of the Spratly archipelago long claimed by Manila. A number of naval vessels had been deployed to the Reef on which the Chinese flag had been planted. China's response that the facilities in question had been constructed to provide shelter for fishing vessels failed to allay regional concerns. The Philippines government registered a formal protest and reinforced its meagre military presence in a token manner.

Despite regional concern about China's creeping irredentism there was a deafening silence from the Philippines' ASEAN partners. Even the United States refused to take sides. On a visit to Malaysia in November 1994, President Jiang Zemin had given assurances that China firmly opposed the use of force or the threat of force in solving disputes, but such assurances evidently did not include the deployment of force against contested, but unoccupied reefs in the South China Sea. Such deployment exposed the weakness of ASEAN's prescription of multilateral security dialogues as a basis for regional order.

An ASEAN initiative to secure the international rehabilitation of Myanmar within a regional framework also met with difficulty. An underlying concern of the Association has been China's ability to profit from Myanmar's international isolation. Chinese technicians have reportedly constructed signals intelligence facilities on Myanmar islands in the Indian Ocean as well as corresponding facilities in southern Laos. Nevertheless, ASEAN pursued a policy of constructive engagement to counter China. Myanmar's Foreign Minister had been invited to attend the annual meeting of ASEAN's foreign ministers in Bangkok in July 1994 as a special guest. But there has been no indication on the part of the military government that it is prepared to modify its draconian rule, symbolised by the continuing imprisonment of Nobel laureate Aung San Suu Kyi. Until it does so, it will continue to be internationally unacceptable for ASEAN to contemplate Myanmar as a candidate member, although individual ASEAN states may consider the kind of 'constructive engagement' that Singapore has been talking about.

China's ties with Myanmar were strengthened in late December 1994 when Prime Minister Li Peng paid a three-day visit to the capital. This was the highest level of formal representation by the People's Republic since the State Law and Order Restoration Council seized power in 1988. At a meeting in Bangkok in January 1995, senior ASEAN government officials expressed concern at the burgeoning military cooperation between China and Myanmar. That Myanmar's army could overrun all rebel Karen strongholds along the Thai border by February 1995 after nearly five decades of resistance testified to the value of the arms transfers it has been receiving from China. The relevance of both China and Myanmar to regional cooperation was demonstrated at a meeting in Hanoi in late November 1994 when representatives from Vietnam, Cambodia, Laos and Thailand initialled a draft agreement for the sustainable development of the Mekong River basin under the aegis of the United Nations Development Fund. The participation of China and Myanmar in the agreement as upstream riparian states is considered to be critical to the full exploitation of the river.

Cambodia Without the UN Transitional Authority

The Cambodian conflict may have been settled as an international problem, but this has not stopped fighting inside the country. Acute tensions remain within the coalition government established in October 1993, but the major confrontation has been between its forces and those of the Khmer Rouge. The Khmer Rouge boycotted the elections conducted under UN auspices, but then sought to secure a place within the coalition government ostensibly in an advisory role. Fighting resumed in late January 1994 when government forces launched a campaign against Anlong Veng, a Khmer Rouge base to the north-west of Phnom Penh, which was captured in early February after its defenders had withdrawn. By the beginning of March, however, the Khmer Rouge had mounted a fierce and effective counter-attack to recapture the base. In the course of this battle the military incompetence and poor leadership of the government forces were clearly demonstrated; it also became clear that the Khmer Rouge had benefited from Thai logistic support.

The pattern of apparent military success followed quickly by actual failure was repeated from the middle of March when government forces captured Pailin, the site of the military headquarters of the Khmer Rouge situated close to the western border with Thailand. By mid-April, Pailin was back in Khmer Rouge hands. Moreover, an advance by a limited contingent of Khmer Rouge forces towards the city of Battambang further exposed the weak discipline and low morale of government soldiers who retreated in disorder.

King Norodom Sihanouk, who was receiving medical treatment for cancer in Beijing, took the initiative to promote a round-table conference on peace and national reconciliation with Khmer Rouge and Phnom Penh government participation in Pyongyang at the end of May 1994. A second meeting took place in Phnom Penh in mid-June, but broke down after the government refused to consider a Khmer Rouge proposal that a cease-fire should only follow negotiations on a power-sharing agreement. The government then ordered the closure of the Khmer Rouge compound in the capital and expelled all Khmer Rouge officials.

The expulsion and a subsequent initiative to outlaw the Khmer Rouge exposed tensions within the coalition government arising from opposition to a proposal by King Sihanouk that a government of national unity be formed under his control. An abortive coup was mounted in early July apparently headed by Prince Norodom Chakrapong, one of King Sihanouk's sons, and General Sin Song, a former Minister of the Interior, both of whom had been dismissed from office the year before after being implicated in an earlier attempted coup. On 7 July, the National Assembly approved legislation which outlawed the Khmer Rouge. Three days later, the Khmer Rouge announced the formation of a provisional government of national union and national salvation led by Khieu Samphan located in northerly Preah Vihear province.

Cambodia's desperate condition was highlighted in the opening sentence of a valedictory report by Australia's Ambassador in Phnom Penh which was leaked to the press. He claimed that 'drift in government, stagnation in the countryside, the army in disarray, the marginalisation of Funcinpec and corruption everywhere favour the growth of insurgency in Cambodia'. Internal struggle was highlighted by the assassination of journalists critical of the government as well as the dismissal in October of the internationally respected Finance Minister, Sam Rainsy, and the resignation in protest of Foreign Minister Prince Norodom Sirivudh, the half-brother of King Sihanouk. The King returned to Cambodia from China in January 1995, seemingly restored to health. But he has not attempted, so far, to change the form of government for one which would enable him to escape the constitutional constraints of his office.

An American military delegation visited Cambodia in mid-September to assess the requirements of the armed forces in the knowledge that any arms transfers carried the risk of illicit sales to the Khmer Rouge by impoverished soldiers. The prospect of American military aid seemed more likely following a visit to Phnom Penh in January 1995 by US Deputy

Secretary of State Strobe Talbott. In November 1994, the First Prime Minister, Prince Norodom Ranariddh, had ruled out a major dry-season offensive against the Khmer Rouge and indicated that his government would instead encourage the defection of their rank and file. A six-month period of amnesty which expired in mid-January 1995, led to the claimed surrender of around 2,000 members of the Khmer Rouge whose effective strength has been reduced to some 8,000.

The Khmer Rouge does not pose a clear and present threat to the government in Phnom Penh, but has continued to demonstrate a capability to harass government forces and, more importantly, to create a climate of fear and despair in the countryside by disrupting rural development and creating a new wave of refugees. It has also sought to sow fear internationally through the abduction and murder of foreign tourists. One of the few bright spots for the central administration in its attempts to contain the Khmer Rouge has been indications that the Thai armed forces have not been as supportive of the Khmer Rouge as they previously had been. By February 1995, however, the Cambodian Army had suffered heavy casualties in a drive against the Khmer Rouge in the north, in Preah Vihear province, despite the earlier assurance by Prince Ranariddh that his government would not embark on an offensive.

Managing Regional Security

The continuing tribulations of Cambodia were raised at the ASEAN Regional Forum in Bangkok in July 1994. The ARF gave no indication that it would be able to take remedial action. Despite the dire condition of Cambodia, however, and the lack of any positive response from the alliance, the difficulties the country faces did not adversely affect any of the bilateral relationships within the ARF. Multilateral security dialogue of a ritualised kind was pursued in the absence of any security issue close to boiling point in South-east Asia. The Korean issue was addressed, although it was not within the legitimate remit of the ARF in the absence of a representative from Pyongyang. The problem of managing power in a world without a common locus of authority hovers spectre-like over the deliberations of APEC and the ARF. Neither of them is competent to address that problem directly which means that the substance of regional security remains elusive. Above all, the ARF and APEC seem unable to address China's creeping assertiveness which has been seen in new intrusions in the Spratly Islands. This was the first direct confrontation between China and a member of ASEAN and arose little over six months after the first working session of the ARF. It makes clear both the need for a regional security organisation and the weaknesses of the present arrangements.

The discussions on security matters that took place in Asia-Pacific during 1994 and early 1995 prove the axiom that for many in the region 'the process is the message'. The regional groupings are still clubs, not alliances; they are pacifiers, but not peacemakers. This is an acceptable arrangement under current conditions, and obviously fits the desires of the regional actors. Whether these new initiatives will survive a genuine crisis in Asian-Pacific security is questionable, however. A strong and consistent American presence, therefore, even if it is somewhat resented, remains vital to ensure the peace that guarantees continued prosperity in the area.

South Asia: Dominated by Internal Politics

Even as the economies of the subcontinent continued to open up to the outside, the politics of the region in 1994 turned distinctly inwards. This contrasted sharply with previous years when the chronic India–Pakistan dispute over Kashmir had dominated the agenda. In several regions the electorates, in a strongly anti-establishment mood, threw out incumbent governments. This was as true in the national capitals of Nepal and Sri Lanka as it was in the provincial capitals of five of the six major Indian states that elected new legislative assemblies. In Pakistan, where no elections were held, rising ethnic and sectarian violence in Sindh province, which includes the Islamic republic's financial capital, Karachi, kept Prime Minister Benazir Bhutto's government hostage to worries over internal developments.

As always, both India and Pakistan blamed each other for their domestic problems and their relations fell to a low unprecedented over the past two decades. There was perhaps less sabre-rattling at the UN and other international fora than there has been in previous years, but in 1994 the two neighbours in effect stopped talking to each other. Pakistan closed its consulate in Bombay on the pretext that India would not give it the building it wanted in the metropolis, and it ordered India to close its mission in Karachi, charging that it was primarily a refuge for saboteurs and spies armed with diplomatic immunity.

The violence in Kashmir receded significantly, although the Vale was still far from peaceful. Under international, and particularly US, pressure Indian security forces violated fewer human rights, and New Delhi suggested that it would soon initiate a plan to improve political life in this troubled province.

Prime Minister P. V. Narasimha Rao's Congress Party, already shaken by internal dissidence, fared badly in India's provincial elections. Foreign-policy issues, in which Rao could actually have claimed some gains, were almost never raised. No one even mentioned Kashmir. Instead the debate centred on corruption, inflation and other domestic concerns.

There was some good news for the region, however, in that overall levels of violence eased. India had a rare year free from any communal rioting and Punjab remained peaceful. In Sri Lanka a new peace process began and a cease-fire with the Tamil Tigers in the north was established. Pakistan's slide out of control was worrying, but only in Afghanistan was there outright war. And while there was hope that *Taleban* (the Islamic Student Movement) might sweep peacefully to power in 1995 having easily asserted itself in the southern Pashtun areas, this – mostly naive – expectation was dashed in March 1995 as *Taleban* came to be seen as just another group vying for power in Kabul. If in Afghanistan the distinction between internal politics and civil war was lost, the big states in South Asia struggled to ensure that the politics of economic growth rather than communal strife remained centre stage.

A Lame-Duck Government in India

The new caste and communal equations that had begun to evolve in 1989 have begun to cast a decisive shadow over India's internal politics. The ruling party suffered major setbacks in several provincial elections because of a combination of indecision and cynical disregard for mounting popular disenchantment with political corruption on the part of the Rao government, and adroit use of populist slogans by some opposition parties. In December 1994, elections were held in Karnataka, India's technology capital which embraces the software export centre Bangalore, and in Andhra Pradesh, Rao's home state. Despite a campaign personally led by Rao, his party was humbled by *Telugu Desam*, a regional party led by fading film star Nandamuri Taraka Rama Rao (popularly known as NTR). With his young new wife in tow (his first had produced 12 children and died several years ago), NTR won over an electorate fed up with Congress Party corruption with the promise of total prohibition and the provision of rice for the poor at 2 rupees per kilo.

Corruption was an even bigger issue in neighbouring Karnataka. The Congress suffered a humiliating defeat in both these southern states, considered to be its traditional bastions. These states had supported Rao when he was challenged by northern leaders such as former Human Resources Development Minister Arjun Singh.

If Rao had any hopes of stemming the rebellion, these were quickly dashed as his party was also defeated in its traditional strongholds of Gujarat and Maharashtra by the right-wing Hindu *Bharatiya Janata* Party (BJP) in elections in March 1995. These two industrial heartland states had benefited most from Rao's economic reforms. Three out of every five dollars invested in India go to these two western coastal states. Of most concern was that the BJP combined with the ultra-right-wing and neo-fascist *Shiv Sena* ('the Army of Shiva') to win in Maharashtra, which includes Bombay. Some reassurance came after the election when the parties promised to move closer to the centre, dropping some of their election rhetoric and coming out clearly in support for deregulation and security for minorities.

As often happens in party politics, successive defeats have brought conflict within the Congress out into the open. The dissidents were emboldened by three other factors. One was the government's ham-fisted treatment of the $2bn securities scandal that rocked India in 1993. Another was a scandal in 1994 over delayed sugar imports at a time of nationwide shortages which pushed up retail prices and led to a loss of $200m to the exchequer. The third was the increasing impatience shown by Rajiv Ghandi's still-influential wife, Sonia, towards Rao for what she saw as his indifference towards the enquiry into her husband's assassination.

Arjun Singh, Rao's deputy in the cabinet and the leading dissident, finally resigned from the cabinet and publicly charged Rao with these issues. Rao hit back by expelling him from the party on disciplinary grounds, taking steps to assuage Sonia Ghandi by speeding up the enquiry, and reshuffling his cabinet to include some well-known Rajiv loyalists. But he tarried too long in expelling two of his cabinet ministers who were tainted with corruption and scandal, and as a result found it impossible to control the damage.

The ethos of the current Congress Party has made it difficult to stem the rebellion. The Party, which grew out of the freedom movement, has become so used to power that it no longer has any ideology. It is held together by the common bond of patronage that power brings. But the moment it appears to the rank and file that the leadership may be incapable of retaining power, dissidence springs up. In the minds of many Congressmen, the countdown to the 1996 general election has already begun and Rao seems an unlikely leader to rally the Party.

But his challengers do not look particularly inspiring either. It is thus widely accepted that India is heading for a spell of coalition rule, with the major question being whether it will be a mix of centrist or left-wing parties. The process of working out new alignments and strategies has already begun, resulting in a lame-duck government. This outcome was highlighted by the colourless national budget for 1994–95 in which the country's leading reformer, Finance Minister Manmohan Singh, avoided any major policy decisions.

Emerging Consensus on National Issues

Surprisingly, the best thing that could have happened to economic reform may have been the electoral drubbing of the Congress Party, which thus far had been seen as the only champion of reform. The Party's defeat showed for the first time that reform now cuts across party lines and has become part of the national agenda. In all four major states where the Congress lost power, new governments showed a zeal for competitive reform, even while spouting populist slogans. Even the BJP, which has made its *swadeshi* (self-reliance) movement a crucial element of its xenophobic electoral platform, promised not to reverse deregulation or oppose foreign investments in Gujarat and Maharashtra which it now controls.

In Andhra Pradesh, the iconoclastic new chief minister passed 23 Memoranda of Understanding for major power projects within 24 hours. H. D. Deve Gowda, the new Chief Minister of Karnataka, continued to spout his left-of-centre socialist rhetoric, but travelled the world soliciting investments. He even promised to ride roughshod over the growing green lobby to entice Dupont to shift an ambitious nylon plant to his state from the neighbouring state of Goa where there was strong environmentalist opposition.

The most extraordinary indicator of this new national consensus was the turnaround in West Bengal where Marxists have ruled without interruption for nearly two decades. The Left Front government, led by the Communist Party of India-Marxist (CPI-M) Chief Minister Jyoti Basu, made a major pitch for foreign investments, deregulating, privatising and, most important of all, promising peace from the trade unions, most of which the party controls. It was no surprise that the Confederation of Indian Industry, the top lobby of the Indian private sector, chose Calcutta for its high-profile centenary celebrations.

The economy demonstrated its new-found resilience, registering export growth in dollar terms of nearly 18% despite almost a month's disruption in international trade and transport because of fears that a plague

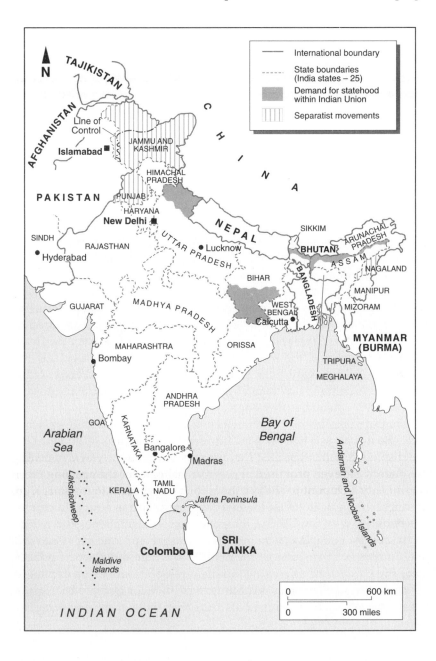

epidemic would spread. Foreign exchange reserves nearly doubled over the previous year's figure of close to $10bn. Industry came out of recession with 8% growth, and the economy grew overall by 5.3%. But these gains were tempered by the return of double-digit inflation as the average annual rate hovered at 10.5–11.5%.

A similar sort of national consensus prevailed in foreign and security policy. While there was some criticism from the opposition of the government's 'soft' attitude towards US pressure on the nuclear issue, all parties backed the basic policy of not signing the NPT and maintaining a hard line towards Pakistan. Prominent opposition leaders were routinely invited to head official delegations to crucial UN and other multilateral meetings. On the fundamentals of economic and foreign policies, India appeared to be speaking with a reasonably uniform voice.

Cosying up to Washington

The most significant foreign-policy event in 1994 was Rao's visit to Washington, from 15 to 21 May, which changed the tone and substance of India–US relations. The visit was preceded by a flurry of diplomatic contacts led on the US side by Deputy Secretary of State Strobe Talbott and Assistant Secretary for South Asia Robin Raphael. These meetings reached a consensus on some troublesome issues. Both sides decided on a two-track approach: to continue talking about the contentious issues quietly, while giving public prominence to the economic dimension.

Newspaper headlines in India therefore drew attention to the rising investments by US corporations, and to two highly publicised visits by Energy Secretary Hazel O'Leary and one by Commerce Secretary Ron Brown. By contrast, there was only muted criticism of the fact that India had not yet deployed the short-range *Prithvi* missile, although this had been scheduled for April 1994. It was obvious that Rao had conceded to Washington's demand not to take any overt action that the US administration could not defend before Congress.

What is crucial about *Prithvi*, which is mounted on a mobile launcher, is that the missile has not yet gone into serial production, although it has been displayed in the Republic Day parades for two successive years. It is now clear that *Agni*, the nuclear-capable intermediate-range missile, is also unlikely to develop beyond the technology-demonstrator stage. India's tactics have been to use the missile programme as a bargaining chip in its dealings with the US. But the policy has its limitations, as anything that is seen as a concession to the US on missile and nuclear issues will have disastrous consequences for popular opinion and election politics. Hence the paradox of displaying missiles on mobile launchers in parades and yet repeating claims (as did Minister of State for Defence Malikarjun at the 1994 Abu Dhabi armaments fair) that these were not being deployed.

Overall, the pressures from the US have reduced a great deal. India has also succeeded in building a lobby in Washington. Thirty top US corporations investing in India have set up an influential India Interest Group that meets periodically. These foreign-policy gains have been strengthened by successful economic diplomacy elsewhere, and by the fact that most of the world has remained neutral on the Kashmir issue.

Rao tried to follow his Washington success with a visit to Moscow where he signed a joint declaration whose significance lay in the inclusion of a statement on the paramount need for the integrity of multi-ethnic and multi-cultural societies as legitimately constituted national entities. Although Prime Minister Viktor Chernomyrdin made a return visit to India, the instability and uncertainty of politics in Moscow imposed limitations on any progress. India and China also moved slowly forward in the process of *rapprochement* during the year. More points of border trade and transit were opened, but the military-level talks on border demarcation and troop disengagement proceeded painfully slowly. India also persisted in its efforts to open up to Iran. The respective foreign ministers exchanged visits and Iranian President Rafsanjani confirmed a new schedule to visit India after an embarrassing cancellation in the autumn of 1994, ostensibly because of the plague scare.

Consolidating in Kashmir

This deft management of foreign-policy interests was accompanied by some tactical gains in Kashmir in a year when relative peace persisted in Punjab, on the communal front and in other chronic trouble spots. The level of militancy was much lower than in previous years. Except for one incident in which irate paramilitary soldiers shot nine civilians after an ambush resulted in the death of their officer, there were no major cases of human-rights violations. More important than the military quiet, however, was the initiation of some sort of a political process.

New Delhi's calculated risk in freeing a majority of the middle-level Kashmir leaders seemed to be paying off, at least in its initial stages. Fulfilling one of their old demands, these political rebels have been allowed to travel abroad. Although none has offered to work for an arrangement short of independence for the troubled state, each has spoken out against violence. Even more importantly, this has opened a channel of communication between the government and the Kashmiris.

The government's new plan for Kashmir is now unfolding. It involves reducing militancy to even lower levels by summer 1995 and then holding elections to the state assembly, hopefully with the participation of the recently freed rebel leaders. This is in keeping with India's experience of insurgencies in the tribal north-eastern states. When the population tired of the fighting that was leading them nowhere, elections brought rebels into power constitutionally and thus back into the political mainstream. In

Kashmir there are limits to this policy. As a precondition, even the most 'reasonable' of leaders are already demanding a degree of autonomy that New Delhi may find impossible to grant. But an election is crucial if any sort of normality is to return in a state which has been under direct central (federal) rule for nearly five years. Pakistan realises what India is trying to do, and is likely to increase pressure by stepping up aid to Indian militants once the snows melt and the passes reopen in summer 1995.

Pakistan: Old Problems in New Dress

Unlike its South Asian neighbours, Pakistan did not face elections in 1994. Yet old problems returned in worrying new manifestations, forcing Islamabad to focus more on domestic than regional concerns. Most important of these was the return of ethnic and sectarian strife in Karachi which resulted in the loss of nearly 2,000 lives. The spurt of violence was sometimes sectarian (Shi'i–Sunni), sometimes ethnic (Sindhi–non-Sindhi), sometimes linguistic (Urdu versus Sindhi-speakers and others) and often a bewildering but bloody mix of all these factors.

The strife that brought Pakistan's financial capital, largest city and only major port to a standstill, attracting international concern – particularly in early March 1995 when two US consulate employees were killed in broad daylight at a busy traffic intersection – has no easy explanation. Like most overpopulated third-world urban centres, Karachi, a teeming city of nearly seven million people, has most of the ingredients of civil strife. People from different ethnic, sectarian and linguistic groups are packed in sprawling slums, often fighting for a slice of a tiny economic cake which simply cannot expand fast enough to satisfy them. The result is violence, particularly as the aftermath of the 1980s Afghan War has brought with it drug mafias and a kalashnikov culture.

Like many other festering problems in the subcontinent, the strife in Karachi is partly rooted in the history of Partition. Millions of Muslims who left India for their new Islamic 'homeland' settled in the urban centres of what was then a sparsely populated Sindh province. These 'refugees' (called Mohajirs) spoke Urdu, which bears no resemblance to the native Sindhi, and were better educated and politically more sophisticated. They were also culturally and ethnically different. With the passage of time they came to dominate urban centres like Karachi and Hyderabad and outnumbered native Sindhis in government jobs and business.

The Sindhis retaliated by supporting nationalistic armed groupings. The Mohajirs found their ambition of sharing more political power blocked by the Punjabis, who are numerically dominant at the national level and who make up almost 90% of the army, and they set up their own neo-fascist political party, *Mohajir Quami Mahaz* (MQM – 'Refugees' Nationalistic Front'). It was led by Altaf Hussain, a charismatic rabble-rouser who now lives in exile in London. At home he faces assorted charges ranging from sedition to robbery. To break the hold of the *MQM*, the Pakistani Army and intelligence agencies engineered a split within the party and nurtured a rival faction called *MQM Haqiqi* ('The Real One'). The result is a free for all between the 'original one' and the 'real one'.

At the same time, an ultra-right-wing Sunni group, *Anjuman-i-Sipah-i-Sahaba* ('Society of the Soldiers of the Prophet') has been targeting Shi'is. The same group is also behind the new phase of fundamentalism persecuting the country's tiny minorities, particularly the Christians. In the midst of this confusion the strength of the drug warlords and mafias has grown to the extent that the government has admitted to difficulty in tackling the challenge of narco-terrorism.

Not only is the government under fire from the ethnic and sectarian violence, but Prime Minister Bhutto has shown neither adroitness nor strong leadership. She managed to weather the initial thrusts from her arch-rival Nawaz Sharif. But she then made the mistake of targeting his family members and party, arresting several of them on flimsy charges. A former minister in Sharif's cabinet was given an eight-year jail sentence for possessing a kalashnikov when scores of these weapons are routinely seen and used to fire volleys of tracers during wedding celebrations. None of this gained Bhutto new friends, and more significantly there are now visible strains in her relationship with the Army.

A new cause of concern is an emerging alliance between Sharif and the exiled *MQM* leader Altaf Hussain, whose following among the Mohajirs is still intact. As is customary in Pakistani politics, Benazir Bhutto has tried to counter these pressures by raising the stakes in Kashmir. The results have at best been mixed. In fact, heightened rhetoric has narrowed her options further *vis-à-vis* her Kashmir and India policies. This was evident when she nominated her titular president to participate in the heads of state summit of the South Asian Association for Regional Cooperation (SAARC) so that she would not have to meet her Indian counterpart.

Yet her overall foreign-policy management has been reasonably clever. Her strategy is to project Pakistan as a modern and moderate Islamic state and to use fundamentalist violence to attract Western sympathy rather than opprobrium. Washington's reaction to the killing of two of its officials in Karachi showed that the strategy is succeeding. Washington's publicly stated position is that Pakistan is a modernising Islamic state, with a prime minister who represents the forces of moderation, facing a threat from religious and drug mafias. But this line has its limitations. The US is unlikely to extend this sympathy to supporting Pakistan's vital foreign-policy interests, particularly with regard to Kashmir. Nor is it likely that the Pressler Amendment that bars all aid to Pakistan because of its nuclear programme would be repealed or diluted.

Sri Lanka: an Optimistic Phase

Following the subcontinental tradition of elections held as the result, and under the shadow, of assassinations and violence, another Sri Lankan government of widows was sworn in. The newly elected Prime Minister (later elected President in a direct election), Chandrika Kumaratunga, is the daughter of former President S. W. R. D. Bhandarnaike, assassinated in the 1950s, and former Prime Minister Sirimavo Bhandarnaike. She was educated at the Sorbonne in Paris and, despite her illustrious political background, rarely participated directly in politics. Chandrika was married to Sri Lankan matinée idol Vijaya Kumaratunga who had founded a minor liberal party. It was only after his assassination in 1990 that Chandrika entered politics. After her successful election as President in September 1994 she appointed an ultra-liberal cabinet, sidelining most of the old stalwarts of her parents' Sri Lanka Freedom Party (SLFP) except her mother, whom she nominated Prime Minister. Under the Sri Lankan constitution the President is the functioning chief executive.

During the first part of 1994 the usual insurgency, terrorist killings and political assassinations continued, most notably that of the former minister and presidential candidate from the United National Party (UNP), Gamini Dissanayake. Yet the changes in Sri Lanka since then have been profound and significant. Wide-ranging policy changes brought in by the liberal new government under President Kumaratunga have aroused new hopes for long-term peace and a possible end to violence.

In the August 1994 general elections the UNP lost power after an uninterrupted reign for almost two decades. This brought in a coalition led by its arch rival the SLFP. The SLFP's new leader, Kumaratunga, immediately sent a negotiating team to the island's troubled northern Jaffna peninsula where the guerrillas of the Liberation Tigers of Tamil Eelam (LTTE) have been virtually ruling unhindered for several years. The new government in Colombo offered to forget the bitter past of ethnic warfare between the Tamils and Sinhalese, and put forward in its place several options involving widespread devolution of powers. Although the LTTE was in no hurry to accept any of the options, the goodwill generated by the negotiations has produced more progress than has been seen since a peace formula brokered by India came unstuck in 1987. A cease-fire was agreed and adhered to more strictly than any in the past. The new government's significant goodwill gesture of lifting the economic blockade on the secessionist region created the climate for further constructive engagement. Kumaratunga also conceded the Tigers' demand for engineers and resources to repair the roads and bridges destroyed during the decade-long fighting.

A vital ingredient of the new mood is that the Sinhalese ultra-left rebels have lost their influence and the new government is remarkably stable. Tourism is increasing and other economic indicators are healthy. If the peace process in the north progresses well, Sri Lanka, which already has the best social indicators in the region, could move away from the region politically and economically. The Sri Lankan elite has for years looked towards ASEAN as a better bet for its future. Last year, analysts in New Delhi were quick to note the growing Sri Lankan impatience with the way SAARC and the South Asian Preferential Trade Association it was supposed to have created have been held to ransom by the India–Pakistan rivalry.

Some uncertainties, however, persist. It was not yet fully clear that the LTTE would abandon its tendency to talk only to buy time and then to strike violently. The government has little to offer the LTTE, which already has virtual sovereignty over the Tamil region. It collects taxes, runs the local government, and has a uniformed police force and army. All it lacks is international recognition. Colombo cannot give that, nor is it likely to opt for a confederal system. The LTTE may be unwilling to settle for little more than they already have in a *de facto* way, but may feel that an assured legally binding autonomous status which lifts the defence burden may be worth a try.

The other uncertainty stems from the trial in the Rajiv Gandhi assassination case which began on 19 January 1994 in the neighbouring Indian city of Madras. Indian prosecutors have accused the LTTE of masterminding the assassination and named its top leadership, including its chief Velupillai Pirabhakaran and intelligence head Potuamman, as key conspirators. If they are convicted, even *in absentia*, the verdict would cast a shadow over the peace process in Sri Lanka as India would then be bound to ask for extradition. The trial could also strain Sri Lanka's newly improving relations with India. For the first time in more than a decade there is almost no contentious issue between the two neighbours. Trade is booming and there is an increasing trend towards Indian corporations investing in Sri Lanka.

Internal Politics and External Relations

The weakening positions of both the Indian and Pakistani governments and increasing attention to domestic politics predictably had a serious impact on relations and policies within the region, and on the region's relations with the West. Rising domestic dissent and political crises constrained the governments' ability to consider any significant shifts in their policies towards each other. Both India and Pakistan avoided contact at higher levels – the last serious round of talks was held between their foreign ministers in February 1994. On the positive side, there were fewer opportunities for clashes at international fora in 1994 than usual, although Pakistan did try to table a resolution critical of India's Kashmir policy at the UN General Assembly, only to withdraw it at the last moment. At a lower level both countries continued their competitive gamesmanship, denying visas to respective diplomats, artists, musicians and journalists and expelling embassy officials.

Some contact between analysts and intellectuals continued, mainly through the US sponsored two-track diplomatic process. But Washington did not get very far in its efforts to forge a consensus on the nuclear issue on the eve of the NPT renewal talks in Geneva. Despite a flurry of diplomatic visits and increasing pressure, Washington was only able to extract minor concessions. India agreed to work to accelerate the move towards a worldwide CTBT, one nuclear issue on which it shares the US view, while refraining from adopting a vocal, ideological position against the NPT.

Pakistan, too, took a hesitant and muted step in the same direction by accepting the offer of US help to set up a modern, new seismic facility near Islamabad. This was to be integrated with the international network under the UN Disarmament Commission's GSETT-3 networking experiment. Some new strains emerged in the Pakistan–US relationship when new evidence traced the New York World Trade Centre bombing in February 1993 to holy warriors graduating from the 'University of *Jihad*' that developed in Pakistan's tribal frontier region during the Afghan War. But both sides were quick to sort these out. At the end of March 1995, Pakistan even asked for help from the United States in closing down these terrorist training camps.

Washington had not reduced its activism in the region. It was generating less controversy, however, because of the new sophistication injected into its diplomacy by the lure of investment. It had had limited success with its efforts to solve two crucial issues, nuclear programmes and reducing tensions over Kashmir. It managed, however, at least partially to rein in the support that Pakistan's intelligence agencies give to Kashmiri militants.

With their own political survival uppermost in the minds of the leaders in the two major nations of the subcontinent, and consolidation of newly acquired power the priority in another, the agenda had turned distinctly inwards. It was no surprise then that neither India nor Pakistan made any major defence purchases, despite some preliminary discussions with Russia by India and with France by Pakistan. Adjusted for inflation their defence budgets were slightly lower in 1994 than in the previous year. These positive trends were likely to persist in the coming year. But far more important was the possibility of a move that could have a profound impact on the region: the Indian bid to hold an election in Kashmir. If India decides to take that risk, its success or failure would have a far-reaching impact on the domestic political equation, and on the future of regional relations.

Africa

At the beginning of 1994 it looked as though things could not possibly get any worse in Africa. A violent war was again raging in Angola after a brief peace, Liberia and Somalia were in the grip of terror and ethnic tensions threatened to explode in other areas. During 1994, however, things did worsen. The most atrocious massacre of one ethnic group by another occurred in Rwanda, new warfare developed in Sudan, Liberia was racked by further vicious fighting and other areas suffered from a complete collapse of civic order. Although a kind of peace returned to Angola, it appeared at best temporary. Only in Mozambique and South Africa was there real hope of a return to normality, a hope that some societies on the continent might find a way to reconcile their once sundered parts and become whole again.

The people of Angola have been told by their politicians so many times that peace has come at last that they may be forgiven for being sceptical about the agreement signed in Lusaka in November 1994. Their doubts were reinforced by the fact that Jonas Savimbi, the leader of the *União Nacional para a Independência Total de Angola* (UNITA), did not sign the Lusaka Protocol himself and that in many respects the war was unresolved. Cease-fires and accords have been agreed many times in the past by one or both sides seeking a respite to re-group and re-arm before returning to the battlefield. Since the agreement, United Nations (UN) and Red Cross personnel have been shot at by UNITA forces and prevented from visiting certain areas.

The Protocol established a cease-fire, the demobilisation of troops, the allocation of four ministries in government to UNITA and the control of three provinces and several cities and zones. The unanswered questions are what role Savimbi is going to play and whether the accord will lead to a *de facto* division of Angola. The process is to be overseen by a UN force of 6,300 troops and 600 observers.

In military terms the government is stronger than it has ever been, but the cease-fire was agreed despite the opposition of the government military commanders who wanted to finish the job. In early October the government army, the FAA, retook the remains of the city of Huambo in the Central Highlands, the provincial capital at the heart of Savimbi's area of support. Much of the city was destroyed by government shelling and bombing which is estimated to have killed some 10,000 people. The government also succeeded in retaking the diamond mining area in the Kwango Valley which UNITA had been using as an important source of revenue.

The crucial factor was the ending of external support for UNITA, particularly from South Africa. Several experienced South African bush fighters, who had previously fought alongside UNITA, signed up as mercenaries to fight for the government. Selling its oil revenues many years in advance, the government bought huge quantities of weapons while UNITA's supplies ran low. Its arms and fuel dumps inside the country and across the border in Zaire could not give it sufficient supplies to hold territory. UNITA had to cash in its chips for political gains or return to the bush as a purely guerrilla army.

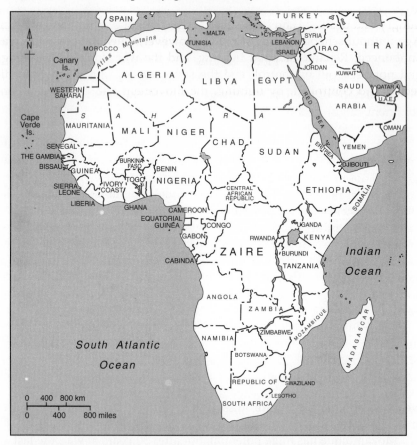

The UN has made it clear to both sides that it will only send and maintain troops in Angola if the country remains peaceful. In diplomatic circles there is optimism and the UN force is oversubscribed. However, there are doubts about the commitment of the United States, which supplied and supported UNITA for many years, to see the peace process through. To spread the financial burden and bind the Angolan state further into the process some of the cost of the UN operation is supposed to be borne by Angolan oil revenues, but few believe that this plan is workable. Meanwhile hundreds of businessmen and traders have descended on the country hoping for a share of Angola's potential wealth from oil, diamonds and agricultural land.

In Mozambique, the most surprising thing about its war was the way it ended. Started by Ian Smith's Rhodesia and then fuelled by South Africa, the war of the *Resistência Nacional Moçambicana* (Renamo) against the Frelimo government of Mozambique seemed to have burst from its bounds. Renamo began as a guerrilla force without social or political ideology, but it generated murderous chaos with gangs of outlaws living by the gun, killing, stealing and destroying as they went. Millions became hungry and destitute and large areas of Mozambique returned to the Iron Age without schools, clinics or roads. However, when the cease-fire was announced, Renamo stopped fighting, and the war stopped. The gangs had not been out of control, and the savagery and chaos had been directed and controlled. By fighting, the movement had obtained a constituency which it would never have achieved by peaceful politics.

In the run up to the UN-supervised election Renamo persistently demanded more concessions, usually more luxuries and privileges for its leaders, by threatening to return to the bush. The last attempt was made when the election was under way and the Renamo leader, Afonso Dhlakama, withdrew alleging conspiracy and fraud. The electors, even Renamo election agents, were ignorant of their party's withdrawal or chose to ignore it and Dhlakama, having secured the attention he wanted, rejoined the process. His party gained 112 seats in parliament to Frelimo's 129 while his personal vote was 34% to President Joaquim Chissano's 53%. Frelimo held the areas close to the capital while Renamo gained a majority in the centre and north of the country.

Bucking the trend in other African countries emerging from long civil wars, the new government has not given any posts to Renamo, even though the former rebels have retired the party faithful and brought in new blood and technocratic elements. Renamo seems to have accepted the role of 'loyal opposition', but the attempt to forge a new army out of the two forces has not yet been a success. Hundreds of thousands of weapons are still hidden throughout the country and millions of mines are planted in small roads and paths.

Mozambique has huge potential, but after 30 years of war, the last 16 particularly destructive, it will take years for the infrastructure to be built or rebuilt. It will also take years to rebuild, or build, in South Africa a society which can merge the long-predominant whites and the long-suppressed blacks. Bringing the country to the point it has now reached has required a miracle; sustaining this miracle will be the difficult task for the future. In most of the rest of Africa, however, to stop the explosions of societies and countries and to begin to reconstruct some semblance of normality will be a major miracle.

Africa's Small but Vicious Wars

Africa's wars were once interpreted as spillovers from the fight against apartheid in South Africa or an extension of worldwide East–West rivalry during the Cold War. Others, who saw through the international and ideological dimensions of conflicts such as Biafra, Angola and Somalia, viewed them as local 'tribal' wars. They are now being seen as something much more significant and sinister. What is happening is the collapse, or more frequently the implosion, of the African states as constituted more than a hundred years ago by the colonial powers.

In many places this implosion is accompanied by the destruction of the economic infrastructure and the rapid disintegration of social systems. Through the bloodshed, chaos and mass movements of peoples, it is not yet clear what will take the place of the states. But, as frightened, vulnerable people take refuge among their kith and kin, there is a reassertion of older, deeper loyalties and hierarchies.

Squeezed from above by pressure from aid donors for less government and forced to cut budgets by the Structural Adjustment Programmes of the World Bank and the International Monetary Fund (IMF), African governments have less power. The organs of state are being dismantled, leaving rulers with less patronage with which to buy support and less power to bind in potential or actual opponents. They also have less cash to pay their soldiers or provide security for their citizens. In some cases, the state can no longer provide schools, hospitals or even roads.

The state is also being eroded from below as the urban and peri-urban masses are severely affected by economic pressure and the mass of subsistence farmers, unable to find markets and decent prices for their goods, lose faith in the state. Both turn increasingly to their own people, people who speak the same language, follow the same customs or belong to the same ethnic group. Loyalties are being tested as never before.

Ever since independence the colonial boundaries which cut across ethnic groups, trading links and other traditional routes and borders, have been questioned. Despite lacking the glue of shared history to hold states together, there have been few attempts to secede or to form new states since the collapse of the Biafran rebellion in 1970. At its inception the Organisation of African Unity asserted its acceptance of the colonial borders. Almost no African leader was prepared to tamper with them or suggest they be redrawn. Even now in states at war there are few explicit attempts to break away from existing states.

But as the central power of states declines, the power of warlords and local barons is growing. That power is regionally or ethnically based and the warlords are fighting, not necessarily for independence, but against the state or the ethnic group which controls it. These movements are, as ever in Africa, difficult to defeat; they live off the land, fight when and where they choose and usually wear down the government forces by destroying infrastructure in guerrilla attacks. Their main weapon is terror. In each of Africa's wars, civilians are the main victims.

One of the most common images of war in Africa is of a gang of scruffy young men armed with AK-47s, in a variety of uniforms, if any at all, emerging from the bush. They might be rebels, they might be government soldiers, or they might be both, sometimes one, sometimes the other. Close on their heels comes a second image: women and children and the old hurrying away from their wrecked and looted homes with babies on their backs and their scant possessions on their heads, desperately seeking safety. The third image is of starvation and disease. Drought or crop failure killed almost no one in Africa last year.

There was war or violent armed and organised rebellion in 26 of Africa's 49 mainland states in 1994. Apart from a spat between Cameroon and Nigeria there were no wars between states. All the wars were within states, though most spread across national borders. Rarely are political negotiations successful in bringing such wars to an end because there is no constitutional context in which to secure and guarantee agreement. At the United Nations the new buzz word is preventive diplomacy and British Prime Minister John Major proposed in Cape Town that the UK should help identify and seek to stop Africa's wars before they start. At present most African leaders trust only their guns.

Rage in Rwanda

Historians are apt to single out two events in Africa in 1994: the election of Nelson Mandela as President of South Africa; and the genocide in Rwanda. Just as part of Africa seemed to be on the edge of a new era of justice and peace, the horrors of the Rwandan nightmare burst on the world. Figures are imprecise, but most observers agree that from April to September 1994 between half a million and a million people were killed in Rwanda.

In the beginning the genocide was planned and ordered. Members of the government of President Juvenal Habyarimana – who was killed with the President of

Burundi on 7 April 1994 when their aircraft was shot down – drew up lists of targets and organised and inspired the gangs of killers. But they cannot have imagined what catastrophe they sparked. At first the gangs hunted down their political opponents and their families and killed them. In the end, throughout the whole country, in every village, gangs of Hutus armed with machetes, clubs and spears hunted down their Tutsi neighbours and slaughtered them. Some of the killings were carried out with mechanistic inevitability – 'it had to be done' – others were carried out with psychopathic brutality. Some victims were chopped at the back of the head, but others were dismembered, or had their feet and hands tied and were thrown into the river, or simply had their feet cut off and were then left to die.

The Rwanda Patriotic Front (RPF), the rebel movement which came to power in July 1994, grew out of a previous pogrom. In 1959, in the lead up to independence, the Hutus, traditionally the peasant farmers ruled by a Tutsi 'aristocracy' in Rwanda and Burundi, rose up and killed thousands of their former masters. Thousands more fled to Zaire and Uganda, where they were left in bleak refugee camps for nearly 30 years. Rwanda became a Hutu state in which Tutsis were discriminated against and constantly humiliated. But the exiles never forgot their home. Their children, many of them soldiers in Ugandan leader Yoweri Museveni's National Resistance Army, formed the RPF and in 1990 launched their attack on Rwanda.

They were beaten back by the Rwandan Army supported by troops from Rwanda's closest allies, France and Belgium. The war settled into stalemate with the RPF holding a narrow strip along the Ugandan border. Peace talks were organised by the UN and Rwanda's neighbours, and eventually, in August 1993, produced the Arusha Accords which provided for a UN peacekeeping force and a power-sharing agreement. President Habyarimana signed, but never implemented, the Accords while building up extremist Hutu elements in his government. These extremists feared the return of the Tutsis, and it was probably they who shot down the President's jet.

Immediately the death squads went into action and anyone thought to be sympathetic to the Peace Accords was killed. The Prime Minister, Agatha Uwilingiyimana, was one of the first to die along with ten Belgian paratroopers guarding her who were ordered by the UN to surrender their weapons and not defend themselves. Despite repeated warnings, the UN bureaucracy and Western governments were quite unprepared for the scale of the massacres.

Meanwhile the RPF launched an attack down the eastern side of the country urged on by the need to stop the massacres of their fellow Tutsi. Fed by government propaganda, the local Hutu people fled before the tide and on 29 April burst through the border into Tanzania in one of the largest movement of refugees ever recorded by the UN High Commissioner for Refugees (UNHCR). The RPF then swung west and tried to surround the capital, Kigali, to link up with the 600 RPF troops there as part of the peace agreement. For weeks they shelled the capital and infiltrated troops into the streets to fight building by building. After three months the capital fell and the RPF announced the formation of a new government. Hutus fled westwards and into Zaire, where many fell victim to cholera and other diseases.

Under pressure from Belgium and the United States, the small UN observer force of 2,500 had been cut to 250 and confined to the centre of Kigali when the killing started in April. Despite pleas from Canadian General Romeo Dallaire, his troops were not permitted to intervene even to help Rwandans murdered in front of them. After long delays, the Security Council in New York decided to send more troops, but before they could move France unilaterally sent in its own troops. The motives for the French intervention are still unclear, although a desire to advance the principle of humanitarian intervention and, as always, to display French 'relevance' to Africa were at least part of the explanation. Certainly, *Operation Turquoise* was strongly opposed by the RPF because France had been the strongest supporter of President Habyarimana's regime. In effect, the intervention helped to stabilise the situation in the west of Rwanda for a while, but it was too late to change the course of the war, or to stop the genocide. Killings went on in the French zone under the noses of the French troops.

There have been many predictions that a similar genocide will break out in Rwanda's neighbour, Burundi, which has the same ethnic mix. But ultimate power in Burundi is retained by the largely Tutsi army and the killings which broke out in 1993 reinforced the army's dominant position. In the face of this, Burundi's fragile politics inch forward, frequently wobbling on the brink of disaster, but not quite falling over.

The Hutu–Tutsi divide now spans not only Rwanda and Burundi, but also eastern Zaire and south-western Uganda, and threatens the stability of the entire region. The defeated Hutu army in the camps in Zaire has not been disarmed and has already launched raids in Rwanda. The appeal by Boutros Boutros-Ghali, the UN Secretary-General, for 5,000 troops for the camps has been ignored. The Zairean Army has made sporadic attacks on the civilian camps in an apparent attempt to drive the refugees back into Rwanda by force, but it cannot, or will not, deal with the camps of the former Rwandan troops.

The former Rwandan Army and the political bosses who instigated the genocide in April dominate the civilian camps around Goma and the international aid agencies that are supposed to control them. In those camps, lying precariously under an active volcano, are nearly a million displaced and bitter people. There is little prospect of them returning to their homes under a peaceful and orderly agreement. In the meantime, the government in Kigali is a largely Tutsi government. Whatever its moral authority, the Tutsis are only about 10% of the population, so any move towards democracy might remove it and perhaps even replace it with the racists who conducted the genocide.

Pulling Out of Somalia

While preventive diplomacy is a sound idea, it also recognises the huge difficulty of intervening once the fighting has started. The post-Cold War dream of a global police force under the control of the UN died and was buried in Somalia on 3 October 1993. The incident began with another attempt by US forces to grab General Mohammed Farah Aideed, the Somali 'warlord' held responsible for the killing of Pakistani UN troops in June. A company of US Rangers and Special Forces entered central Mogadishu and were trapped. Two helicopter gunships went down and the rescue force was ambushed. In the shambles that followed, 18 American servicemen died and the US Special Forces had to be rescued by Malaysian troops.

The sight of a dead US serviceman being dragged through the streets of Mogadishu was unbearable for America. The deaths of those soldiers had huge implications for US policy. Washington reacted by withdrawing from Somalia shortly thereafter. Washington then argued in the Security Council for the withdrawal of the whole UN force from Somalia and President Clinton signed Presidential Decision Directive 25 curtailing future US commitment to international peacekeeping. The real victims of the Somali war were the Rwandans and Bosnians.

The abandonment of Somalia by the UN at the end of March 1995 was almost like the defeat of an imperialist power. Somalis who once welcomed the UN troops were jubilant at their departure. The departure itself was a huge operation involving the warships of three navies, helicopters and amphibious landing craft, and was carried out smoothly with no loss of UN life.

For General Aideed and his rival Ali Mahdi Mohammed the departure of the UN leaves a vacuum that only one of them can fill. The prizes that the UN leaves behind are the airport and the sea port as well as the huge UN compound erected at great expense in the strategically placed US embassy. For the second time this decade it will be comprehensively looted down to the wiring in the walls.

The UN had become irrelevant in Somalia's six-year-old war, and its departure may not change the political and military position on the ground. In southern Somalia the clans are no nearer a peaceful solution than they were when the war started. Skirmishes between clans have continued almost daily in Mogadishu and fighting is regularly reported from the Belet Weyne area, Kismayo, Medina and many other areas disputed by the clan-based militias. Their alliances are constantly shifting as the militias attempt to grab strategic areas or simply seek revenge for a previous attack. In other parts of the country, Somalis have learnt to exist without government and are finding ways of living, producing and trading within informal structures.

General Aideed remains the most heavily armed of all the leaders, but his Habr Gedir sub-clan of the Hawiye clan is not powerful enough to take over the whole country or even the whole capital. The more he tries to dominate, the more he creates resentment in the other clans. His main rival in Mogadishu is the Abgal clan of Ali Mahdi which controls the north of the city. Both Aideed and Ali Mahdi held peace conferences last year in attempts to keep their clan alliances intact, but there is little chance of the two establishing a lasting peace. The stand-off is expected to continue unless Aideed provokes another battle by advancing territorially or announcing a national government.

The outside world has lost interest in Somalia. With the departure of the UN the only outsiders with a stake in the region are the Eritreans and Ethiopians who have geopolitical reasons for wanting stability. Meanwhile arms and ammunition have been pouring in by sea and across the Ethiopian border. The best that southern Somalia can look forward to are sporadic cease-fires.

In the north the hopes awakened by the secession of Somaliland have been dashed by clan warfare following the collapse of the Somali National Movement, the rebel movement which formed a government and declared independence in 1991. For most of last year, the self-proclaimed government of elders battled with clan militias for control of Hargeisa airport and the port of Berbera, the country's vital links with the outside world. Prospects for the reunification of Somalia are not good.

Sudan

The Horn of Africa is so far the only region on the continent where states have formally broken up. Sudan is almost certainly the next candidate and there is little hope that it will ever come together again as a unified state. Twelve years of war between the largely Christian or animist black African south and the Islamic Arabic-speaking north have left about a million dead, mainly from hunger and disease, and about two million displaced.

The government in Khartoum, heavily influenced by Islamic fundamentalism, seems determined to reassert its military authority over the south and last year took further advantage of the split in the southern rebel movement, the Sudan People's Liberation Army (SPLA). The two wings of the movement broadly based on the Dinka and Nuer tribes, turned on each other after John Garang's deputy, Riak Machar, tried to overthrow him in 1991. Garang had already been weakened by the collapse of the Mengistu regime in Ethiopia which had kept the SPLA supplied and provided with a rear base. During the last two years the two tribal groups attacked each other's civilian population with ferocious barbarity.

Prior to the split the movement controlled all the south with the exception of Juba and a few besieged garrisons. Now the government forces have regained almost all the territory lost in the 12 years of war and at the beginning of last year they were poised to push the remnants of the SPLA over the border into Uganda. In February the government launched an offensive against the SPLA's stronghold around Nimule and forced thousands of civilians, the displaced families of the Dinka fighters, to retreat to the border. The government armies advanced south in three columns and conducted haphazard, but terrifying bombing raids on civilian centres from the air. The SPLA, lacking guns and ammunition, seemed on the verge of defeat, while the Sudanese Army began to arm rebel movements in Uganda as a way of destabilising the SPLA's source of supply.

In March, however, convoys of weapons were seen crossing into Sudan at Lokichoggio in northern Kenya and suddenly John Garang's fighters were re-armed and reorganised. The resupply continued throughout the year. The source of the weapons is unclear, but Israel and the United States remain the leading suspects.

Sporadic peace talks had gone on throughout most of the year, but as the dry season approached in October they collapsed and both sides returned to the battle front. The re-armed semi-guerrilla army of John Garang has launched attacks on five fronts and is expected to do well against the Sudanese Army which can easily be pinned down in towns and ambushed on the roads. Garang's rival, Machar, has given up his struggle for the leadership of the movement, and in early February he declared a cease-fire and reconciliation with the SPLA.

Garang has always proclaimed support for a secular democratic Sudan while Machar has called for independence for the south. Garang now seems to have come round to this view too, even though he realises it may cost him support among the many northerners who oppose the government in Khartoum, but do not want Sudan broken up. If the cease-fire between Garang and Machar's forces sticks, it will stop the tribal killings which engulfed parts of the south and will enable Garang to turn all his energies on the Khartoum forces. There are continuing threats from two of Machar's former lieutenants, however, who are believed to be working for Khartoum and who are still at large in the south-east.

Despite painful economic hardship in Sudan, the government is safe from popular discontent, vulnerable only to a coup. It maintains a vigorous secret police and pays little regard to human rights in its search for plots and traitors. In the long run this creates more opposition, and recently a 'Declaration of Political Agreement' was drawn up in Eritrea between all the main opposition movements north and south. Details have not yet been released. All previous attempts to form a nationwide opposition movement in exile have failed. Sudan may soon find itself isolated in the region. Eritrea, which owes its existence partly to Sudanese support for its liberation struggle in the 1970s and 1980s, has now fallen out with Khartoum and broken off relations. Relations with Ethiopia are also strained and soon Sudan may find itself surrounded by enemies who are prepared to succour rebel movements.

Viciousness in Liberia

Since the Biafran war ended more than 20 years ago West Africa had been spared the horrors that had been witnessed in Uganda, Angola and Mozambique. There has been political upheaval and widespread repression, but West Africans seemed to have little enthusiasm for the long, drawn-out guerrilla war. And then Liberia fell apart. After five years, that war has not only continued with appalling and often bizarre atrocities, but it seems now to have spread to Sierra Leone and may indicate a future pattern for the region.

The failure of a negotiated settlement in Liberia has, until now, been due to the refusal of the Nigerian-led ECOMOG, the peacekeeping force of the regional forum, the Economic Community of West African States (ECOWAS), to allow Charles Taylor, leader of the National Patriotic Front of Liberia (NPFL), a share of political power. He, in turn, has failed to fulfil agreed commitments. Recently there has been evidence that the United States and Nigeria have come to accept that there is little chance of a peace settlement without Taylor.

It was Taylor's NPFL which began the war against President Samuel Doe in 1989. Although Doe was overthrown, Taylor was unable to snatch the fruits of victory. Other ethnically based movements, which received the backing of ECOMOG, sprang up as the Doe regime collapsed. Its 10,000 Nigerian troops have held the capital, Monrovia, since 1993 and protected the interim government set up by the Cotonou Peace Agreement and its leader President Amos Sawyer.

Sawyer's rule extends not further than the outskirts of the capital. The south-east of Liberia is held by the Liberian Peace Council (LPC), the essentially Krahn remnant of President Doe's army. The North of the country is dominated by the United Liberation Movement (ULIMO), which has a Mandingo faction and a Krahn faction. The north and east, perhaps two-thirds of the country, are held by the NPFL. There are myriad other militias and movements.

When the mandate of the transitional government ran out in September 1994, President Jerry Rawlings of Ghana called the leaders of ULIMO, the LPC and the NPFL together at Akasombo on Lake Volta where they agreed to slice the cake between them. A council of five – the three leaders of the armed factions plus two nominated civilians – was set up with the idea that it would be given authority to run the country until elections were arranged for November 1995. This cynical plan to carve up the country among the warlords was endorsed by ECOMOG, but rejected by the civilian assembly in Monrovia.

Meanwhile, a faction of Taylor's erstwhile supporters saw their chance during his absence in Ghana and seized his headquarters at Gbarnga. In Monrovia a former security chief in Doe's government, General Julue, took over the presidential palace and had to be blasted out by ECOMOG troops. Since then Taylor has fought back and thousands of civilians have been forced to flee their homes again. A second meeting at Aksombo has reaffirmed the agreement to set up a five-person Executive Council and it seems likely that Taylor will be given the chairmanship, a step away from the presidency he has coveted for so long.

The war in neighbouring Sierra Leone dates back to Taylor's desire to attack ECOMOG's base there. Taylor lent his support to the disaffected former Sierra Leonean soldier, Foday Sankoh, in 1991. His movement, the Revolutionary United Front (RUF), raided from across the Liberian border, which sparked the civil war. Now it has spread all over the country and only the Freetown peninsula is still secure in government hands.

The RUF claims to stand for a government of national unity and the expulsion of all 'foreign troops' from Sierra Leone, meaning ECOMOG's Nigerian soldiers. The government of President Joseph Momoh was overthrown in 1993 by a group of junior officers led by Captain Valentin Strasser, but the new regime believed it could win the war militarily. The war intensified and the Sierre Leonean Army more than doubled in size.

Many diplomats believe that the original Sankoh is dead or has been replaced by someone who has adopted his name. Others believe that the rebel movement barely exists, and that most attacks are carried out by bandits or undisciplined government soldiers. The attacks have spread steadily westwards from the Liberian border causing thousands to flee their homes. Some 50,000 have crossed the border into Guinea.

Like the Liberian war, this one is characterised by sudden attacks by armed gangs who emerge from the bush to destroy villages, ambush roads and murder, rape and steal. No one knows who they are. Usually they are in uniform, but do little to politicise the villagers they attack or leave any impression except terror. Nor have the rebels made much effort to put their message across to the outside world; they have only a semi-literate spokesman in Abidjan. Yet the taking of foreign hostages who are then kept in safety suggests a political brain of some sort behind the movement.

No Succour in Sight

Predicting Africa's wars is never easy. The potential for sudden outbursts of violence exists in most countries as rising populations meet falling living standards, and weak governments confront regional or ethnic movements. The loss of superpower sponsors means that African governments can no longer fall back on a arms supplies and assistance to suppress their subjects, but a neighbouring state is usually willing to back a rebel movement.

Aid levels from the Western world will fall again this year and there is no evidence that Western democracies are prepared to offer appropriate assistance, even to those countries that fulfil their demands for democracy. The failure of outsiders to help is illustrated by the overthrow of Sir Dawda Jawara, President of Gambia for 30 years. He was deposed in July by a handful of soldiers who had intended only to protest about pay, but found they had taken over a country. A US naval vessel in Banjul harbour gave Sir Dawda a lift to Senegal but did not intervene, while Britain took several weeks to suspend aid to the new government.

Despite appeals for democracy still emanating from London and Washington (Paris is more committed to stability than democracy in Africa, although it has put pressure on some countries, such as Chad, to join the democratic bandwagon) the military rulers who now dominate Anglophone West Africa know they have nothing to fear. In Ethiopia, however, the experiment of trying to establish a federation of nations from Ethiopia's old empire may prevent its dissolution, and President Museveni's attempt to create an inclusive political system with no parties may mark a genuine departure from Africa's repressive winner-takes-all politics. But both experiments could collapse in bloodshed at any moment.

The main thrust of rebellion in Africa is now regional and ethnic. Conditions exist in Chad, Kenya, Senegal, Nigeria, Zaire and several other countries for such rebellion to break out at any time. Once war starts it takes on a momentum of its own and rapidly degenerates into appalling and meaningless atrocity and terror. Although some movements may establish a code of disciplined conduct towards civilians and seize the moral high ground, there is no ideological element in Africa's wars. The only universal element which has the potential to spread is Islam, either as a code of morality and political order or as a violent and exclusive dogma. In West Africa, the Sahelian states and the east coast of Africa, the rise of a violent brand of Islam will continue. In January 1995 Iran began broadcasting to Africa in Hausa as well as in Arabic.

The greatest threat comes from discontented youths who, whatever their original motives for picking up a gun, become caught up in destructive, aimless and endless wars. The targets of these wars are the few wealth-producing centres left in Africa. The victims are the civilian populations.

South Africa – Can the Miracle be Sustained?

After 300 years of oppression, 46 years of apartheid and four years of negotiation, suddenly it was all over. For three days in April 1994 South Africans of all colours and persuasions, black and white, Indian, 'Coloured', English-speaker and Afrikaner came together to vote in the country's first free and fully democratic elections. In doing so they did more than close the book on their past; they also defeated the widespread expectations of a watching world that South Africa would not be able to escape its manifest destiny – an ethnic blood-bath. Instead, as they waited patiently in long queues to cast their ballot, the people of this long – and bitterly – divided country seemed to find both each other and the peace that had for so long eluded them.

If the new South Africa was able to deliver racial reconciliation, however, it failed in its first year to deliver the reconstruction which was the other equally pressing half of the equation. Despite the promises of the ambitious Reconstruction and Development Programme (RDP), the centrepiece of African National Congress (ANC) policy outlined by President Nelson Mandela in his first speech to parliament in May 1994, few houses were built, health services remained chaotic, education in the townships was near collapse, law and order was under threat, and corruption had begun to stain the new government. If the major beneficiary of South Africa's transformation was a rapidly expanding black bourgeoisie, the millions in the squatter camps complained that liberation from oppression had brought no improvement to their lot. Whether South Africa will be able to do so quickly enough will determine whether the brave renewal continues or is halted by a recurrence of uncontrollable violence.

Establishing a New State

The political violence that had the country on the rack stopped almost completely during the election. The bomb blasts that only a few days earlier had claimed yet more lives and sharpened fears of a white backlash proved to be the extreme right's final spasm. The threat of outright civil war between the ANC and the Zulu-based Inkatha Freedom Party (IFP) was similarly averted. At the beginning of April thousands of spear-carrying Zulus marching in protest through the streets of downtown Johannesburg were greeted with a fusillade of bullets from security guards at the ANC's headquarters. When the smoke cleared 53 lay dead and more than 400 were injured. As the Zulu heartland of KwaZulu-Natal was gripped by a spate of revenge killings, President F. W. de Klerk declared a state of emergency in the region, ordered in the troops and insisted that the election would nevertheless continue as planned.

Just days before the country went to the polls IFP leader Chief Mangosuthu Buthelezi was persuaded in a frantic series of last-minute meetings with ANC leader Mandela and the outgoing President de Klerk to abandon his boycott of the election. It was the last hurdle. The election, far from being a trigger for conflict and revenge, became instead, for whites as well as blacks, a strange ceremony of release and reconciliation. Despite a process that had taken three years to negotiate and after a bitterly contested and often bloody campaign, the very act of voting seemed more important than the result.

The Independent Electoral Commission (IEC), its already chaotic organisation further bedevilled by the IFP's late entry, failed to cope with the task in hand and in the end the result of the most closely observed election in history was decided more by negotiation than democracy. The ANC just failed to achieve the two-thirds national majority that would have enabled it effectively to write South Africa's final constitution on its own, and the IFP secured the clear majority in KwaZulu-Natal. Further conflict was thus avoided and the IEC as well as the army of international monitors declared the election to have been reasonably free and fair.

Of the seven political parties to win seats in the National Assembly, three – the ANC with 252 seats, the National Party (82) and the IFP (43) – were large enough to join together to form the government of National unity (GNU) which will govern South Africa until 1999. F. W. de Klerk became a deputy president and Chief Buthelezi Minister of Home Affairs. The right-wing Freedom Front (9 votes), the radical left Pan-Africanist Congress (5) and the Liberal Democratic Party (7) were all but obliterated under the proportional representation system. The ANC also won control of seven of the nine new provinces, with KwaZulu-Natal going to the IFP and the National Party winning the Western Cape, where fears of black domination kept the Coloured majority away from the ANC.

On 8 May 1994, some 40 world leaders, a hundred thousand South Africans and an international television audience of billions watched Nelson Mandela, South Africa's first black president, take the oath of office in front of the Union Buildings, long the seat of Afrikaner power and white privilege. As a South African Air Force squadron roared overhead trailing the colours of the new flag, the cheers of the crowd expressed a new pride of ownership from people who only a month before had regarded the South African military as an instrument of oppression.

International admiration of the South African miracle and its chief miracle worker, President Mandela, led to his receiving a Nobel Prize, shared with de Klerk, and invitations to address the UN General Assembly and the US Congress. Within weeks of the election South Africa was admitted to the Organisation of African Unity, the Southern African Development Community, the Commonwealth and the non-aligned movement and once again took its seat in the UN General Assembly. Diplomatic relations with foreign countries expanded from 36 in 1990 to 124, and the former Olympic outcast put in a strong bid to host the Games in 2004. President Mandela's determination not to reject such old ANC friends as Cuba and Libya may have raised some eyebrows in Washington, but did nothing to temper the country's welcome back into the world community. It was also expected that the most powerful country in Africa would begin to help out with the continent's manifold problems, but the government resisted US pressure to send soldiers to Rwanda, although President Mandela was able to bring his personal authority to bear in brokering a settlement in Mozambique and in the search for peace in Angola.

The spirit of reconciliation between the races which had won President Mandela the admiration of the world persisted throughout the year following the election. The Afrikaner extreme right remained quiescent and increasingly marginalised. Sporadic violence in KwaZulu-Natal rumbled on and Chief Buthelezi walked out of parliament – and walked back in. The National Party, under a strangely detached de Klerk, chafed at the ambiguity of its new role as both opposition and junior partner in government. But the racial sting had been largely removed from South African politics, to the point where issues such as the protection of minorities, which had so bedevilled the multiparty talks, seemed within months of the election to be oddly anachronistic. South Africa's history continued in some ways to haunt its present with the continued disclosure of former atrocities by secret agents of the apartheid state and even more secret amnesty deals, but the main instrument to exorcise these ghosts, the Truth and Reconciliation Commission, was slow to get off the ground.

For the rest, there was no great renaming of places. The industrial heartland of South Africa, the Pretoria–Witwatersrand–Verenigeng area, became the new Province of Gauteng (gold). But in parliament, portraits of the architects of apartheid still adorned the lobbies, although the old House of Assembly where Verwoerd, Vorster and Botha had once ruled supreme became instead the ANC caucus room. Afrikaans remained one of the 11 official languages, with English pre-eminent. *Die Stem* was retained as the National Anthem alongside *N'kosi Sikelele Africa* and, at the start of the 1995 academic year, white state schools across the country were fully and peaceably integrated.

The Dark Side of the Coin

Within months, however, the government was struggling to assert its authority over black nurses, students, teachers, police, prison officers and convicts, who demanded that South Africa's black leadership begin to dispense instant rewards to its followers. In pursuit of a series of demands

– ranging from instant promotion, to higher pay, to amnesty for convicted criminals – lorries were hijacked, universities and prisons vandalised, superiors taken hostage and police stations occupied and besieged. Finally, at the opening of the 1995 parliamentary session, Mandela warned that he would crack down on strike violence, dissent in the police and the army and on any threat to the country's stability. 'The battle against the forces of anarchy and chaos has been joined', he said. The warnings may have allayed the fears of nervous investors, but did little to quell dissent, because the government seemed unable or unwilling to give practical effect to Mandela's words.

The Defence Force embodied the persistent contradictions of the new South Africa. Renamed the South African National Defence Force (SANDF), it seemed on the surface to be one of the first successes of the new spirit of reconciliation. Its generals were content with the appointment of the hawkish Joe Modise, former head of the ANC's armed wing (*Umkhonto we Sizwe*) as new Minister of Defence. Modise, who reappointed General Georg Meiring as Defence Force Chief, rapidly became a key defender of South Africa's war machine and its budget. Despite a pressing need to exact a peace dividend from the military to help fund its RDP, Modise and his generals stood firm against major cuts. In its first budget, the government increased its allocation to 'security' needs – an allocation which, at $700 million, equalled the entire reconstruction fund. One reason was the generals' expensive tank and aircraft shopping list; another was the agreement to absorb and re-train former guerrillas as well as the soldiers from the former 'homelands' into the Defence Force. Under this plan the SANDF would swell to about 120,000 before being steadily reduced over three years to current force levels of around 91,000.

The scheme, however, ran into almost immediate problems as the former guerrillas, chafing under strict military discipline, staged a series of strikes and mutinies that ended with the summary dismissal of more than 2,000 on the instructions of an exasperated President Mandela. Unrest, however, continued well into 1995 when soldiers of 21 Battalion staged a protest march to the Union Buildings.

The erosion of law and order did not diminish South Africa's attraction for illegal immigrants and economic refugees from the rest of the continent – a development which could strain regional relations in the future. By mid-1994, illegal immigrants were said to be crossing the border at the rate of one every ten minutes, and by the end of the year estimates ranged from two million to eight million, of whom only 91,000 were deported.

Open borders did more than drain already strained resources, they also turned the country into a tempting target for international crime. If political violence had dropped dramatically in the aftermath of the election (despite persistent feuding in KwaZulu-Natal), criminal violence, which gave South Africa a murder rate seven times that of the United States, showed no sign of a similar decline. Taxi wars between rival companies fighting over lucrative 'minibus' routes claimed the lives of several hundred people in Johannesburg and Cape Town in two months and were only one more proof that South Africa remained a very violent society. Of greater concern to the West – and particularly the United States, which opened an FBI office in Johannesburg – was the country's growing reputation as a major market for the international drugs trade, with Nigerian and South American gangs, as well as the Chinese Triads, all reputed to be active.

Economic Disabilities

While South Africa attracted the attention of international criminals, conventional investors were more wary of potential instability and criminal violence, as well as the comparatively high wages and low productivity which put the country at a major disadvantage *vis-à-vis* its East Asian competitors. What the World Bank and Japanese Exim Bank, the European Union and the United States were willing to offer was aid in the form of soft loans; but the South African government, committing itself to a monetary and fiscal discipline almost unique in Africa, was reluctant to accept. By the time of the 1995 budget, this probity had begun to achieve results.

Although the predicted flood of foreign investment had yet to materialise, most American and many European companies which had fled during the sanctions era had begun to return. Expected economic growth of 3.5% for 1995 was still far short of the 8% needed for South Africa to begin to dent its unemployment rate of 30–50% of the economically active population. Nevertheless, tourism was growing fast, agricultural exports were flourishing and Chris Liebenberg, the Minister of Finance, was sufficiently emboldened by the revival to announce the abolition of exchange control for non-residents. Against expectations, the rand responded by strengthening against the US dollar. Liebenberg (a former banker and a non-political appointment) followed this with a prudent budget which, although doubling the allocation to the RDP to R5 billion (US$1.4bn), also revealed a new lower target budget deficit of 5.8% of gross domestic product and a government commitment to a public-sector pay increase of only 3.25% against an inflation rate of 9.9%.

This decision seemed set to put the government on a collision course with the trade union movement, a powerful segment of the ANC's support base. There is little doubt that the need for fiscal restraint imposed severe constraints on the government's ability to produce immediate economic dividends for its supporters. It was, nevertheless, widely acknowledged as essential if the country was to retain and attract the investment necessary for growth.

The initial failure of the RDP to deliver even incremental improvement in the lives of the poor appeared, however, to owe less to financial constraints and more to the culture of protest which had grown up during the apartheid years and which the new government was finding impossible to dispel. Plead as he might, President Mandela failed to persuade tenants and householders in the black townships that they would have to abandon their boycott of mortgage repayments, rents and service levies before banks and building societies could be persuaded to guarantee funds for the house-building programme.

The Difficult Legacy

Similar interventions by the President, whether insisting that soldiers return to their barracks, that school children return to classrooms or promising that corruption would have no place in the new South Africa, went largely unheeded. His exhortations were frequently undermined by populist politicians, particularly in the provinces, who seemed unable to withstand the temptation to make common cause with their more radical supporters. The problem went to the heart of the ANC's dilemma: its unwillingness to transform itself from a broad protest and liberation movement into a political party with all the responsibilities of government.

To do so would involve a loss of support to more radical movements and the desire to maintain unity at all costs resulted in an insistence on the widest possible consensus before the most elementary decisions were taken. This in turn slowed the processes of government. A proliferating number of select committees, each of 40 members or more, provided parliament with new oversight of every ministerial portfolio and, in the first year, very little legislation emerged from the process. The Land Act, which will address the claims of the many thousands dispossessed by apartheid, was passed. The Constitutional Court, the ultimate guardian of the Bill of Rights, was established and the Chairman of the Constitutional Assembly, Cyril Ramaphosa (who had lost out to Thabo Mbeki as Deputy President) launched a two-year process which will result in South Africa's final constitution. An initial determination to invite and hear all submissions no matter how frivolous, however, seemed likely to ensure that it would not meet its deadline.

In the meantime, the provincial governments, especially those led by the ANC which had previously opposed a federal constitution, began to exert and revel in a considerable degree of local power. Chief Buthelezi, however, fearful of ceding this power to the centre in a future constitution, continued to press for a fully autonomous KwaZulu-Natal which would, in effect, be the reincarnation of the pre-colonial Zulu Kingdom. Just before the elections, King Goodwill Zwelithini had withdrawn his support for Buthelezi and transferred his allegiance to the government. Among the

Zulus, pressure was growing to force the King to abdicate in favour of a more pliant member of the Royal family.

A major achievement in the restructuring process was the creation across the country of Transitional Municipal Councils (TMCs), which joined together black local authorities and much wealthier white councils in single municipalities, now mostly under black control. These TMCs were set to govern until the first local elections on 1 November 1995, but a degree of disillusionment was reflected in a very low voter registration – astonishing in a country where only a year before 80–90% of the electorate had turned out for their first experience of democracy.

This disillusionment had a great deal to do with the government's failure to deliver the 'one million houses by 1999' it had undertaken to provide under the RDP, although the most effective member of the cabinet, until his untimely death on 6 January 1995, was the Minister of Housing and former Communist Party chief Joe Slovo. Despite the persistent township culture of non-payment, and open hostility from politicians who had promised their followers finished housing at minimum cost, Slovo eventually put together a realistic plan that appeared both workable and achievable and would have begun to offer an incremental improvement in the lives of the poor. Its results would only be seen in the future, however. By the end of 1994 the country's vast housing backlog remained stubbornly untouched, while the growing squatter communities became ever more impatient and frustrated.

The major beneficiaries of the new South Africa were not the homeless or the unemployed, but the rapidly growing ranks of the black bourgeoisie, who found themselves ardently courted by both the private and the public sectors as society tried through affirmative action to change the colour of privilege. Although the top-to-middle ranks of the civil service remained predominantly white, many Afrikaner bureaucrats were tempted by the extremely generous pensions and redundancy payments, negotiated by the previous regime, to vacate their posts to black colleagues. The cabinet also began to employ an ever-growing army of 'super bureaucrats' – political advisers and consultants to monitor and offset if necessary the advice of the 'old' civil service.

Hanging in the Balance

These were the unavoidable signs and symptoms of one of the most difficult, if peaceful, transitions in contemporary history. Yet, they all contributed to a sense of sluggishness in government and the concomitant inability to give effect to its decisions or to control the instruments of power. As politicians began to rely on patronage to win and retain support, the taint of corruption spread through political ranks, with some famous names facing allegations relating to the misuse of power and funds. The most well known – and the most difficult to deal with – was

Winnie Mandela, the President's estranged wife. A flamboyant icon of the radical youth and the poor, Mrs Mandela had, ever since her trial on charges of kidnapping in the case of the murdered youth Stompie Moeketsi, successfully defied every attempt by the moderate ANC leadership to discipline or control her. With allegations of the misuse of funds swirling around her and her reputation permanently stained by the memory of Stompie's death, her dictatorial style provoked walk-outs by fellow executive members of the powerful ANC Women's League. Her popularity with her radical constituency nevertheless appeared to feed on every bungled attempt to bring her down. Eventually, she came to be regarded by her cowed ministerial colleagues as an Evita Peron in the making, relying on a heady mixture of patronage, fiery populist rhetoric, glamour and defiance to shield her from her enemies and to maintain her popularity with the masses.

For President Mandela, the threat posed by his wayward wife was a personal as well as a political tragedy. The way in which he was finally driven to deal with it was instructive. As her presence in the government as Deputy Minister of Arts, Culture and Science, became more and more embarrassing, the President initially relied on his deputy, Thabo Mbeki, to bring her to book. As her repeated attacks on government grew increasingly strident and she deliberately flouted Mandela's authority by ignoring his order not to leave the country for a meeting abroad, he was compelled to take action himself. On 27 March 1995 he relieved her of her post.

Whether or not the amazingly resilient Mrs Mandela was more dangerous outside government than in, whether or not she would be able to make good her threat to mobilise the masses to her cause, there seemed no other choice that would not harm the government even more. What was of concern was that the episode illustrated, once again, that the government had to rely on one brave and resolute but increasingly frail old man for decisiveness and action.

This possibly seminal event took place two days after a week-long State visit by Queen Elizabeth in which the British monarch invested President Mandela with the Order of Merit and had lauded the transformation of South Africa as a miracle. The Royal visit and the sacking of Winnie Mandela encapsulated an extraordinary year for South Africa. If Nelson Mandela offered reconciliation to South Africa's people, his wife promised only retribution; if he was the country's hope for the future, Mrs Mandela embodied its worst nightmare – a descent into the quick-fix policies and demagogic style which have beggared much of the rest of Africa. A year after Mandela's inauguration as President, South Africa had achieved a miraculous escape from its past; as it struggles to come to terms with its future it will need more than the efforts of one man to ensure that the miracle endures.

Chronologies

US and Canada

January

1   North American Free Trade Agreement (NAFTA) enters into force, creating free-trade area between the US, Canada and Mexico.

18–   US Defense Secretary nominee Bobby Ray Inman withdraws; President Bill Clinton names William Perry Defense Secretary (24).

25   In his State of the Union address, Clinton promises reform of health care, and no more cuts in defense expenditure.

31   Gerry Adams, leader of Sinn Fein, the political wing of the Irish Republican Army (IRA), begins 48-hour visit to US.

February

7   Clinton proposes 1995 budget, with a 1.3% rise in defence spending for 1995, but decreases thereafter.

21–   Aldrich Ames, senior CIA official, and his wife arrested on charges of spying for the Soviet Union and Russia; US expels counsellor at Russian Embassy (25).

March

3   US government reactivates 'Super 301' trade law as trade tensions with Japan rise.

4   Ukrainian President Leonid Kravchuk visits the US, which doubles its aid to Ukraine to around $700m a year.

5   US White House legal counsel, Bernard Nessbaum, resigns over Whitewater scandal.

11–   US House of Representatives approves Clinton administration's 1995 budget; Senate follows suit (25).

April

22   Former US President Richard Nixon dies.

28   Aldrich Ames convicted of spying for the Soviet Union and Russia.

May

3   Clinton orders overhaul of US counter-intelligence operations, following Ames spy scandal.

11–   US Senate backs lifting arms embargo on former Yugoslavia; Clinton opposes Senate vote (25).

June

9   US government announces it will pay compensation to veterans suffering from 'Persian Gulf syndrome'.

9   US House of Representatives votes to allow arms shipments to Bosnia.

20   Canadian government restores aid programmes to Cuba.

27   Clinton reshuffles White House staff; David Gergen becomes Special Adviser on Foreign Affairs.

July

18   Canadian Prime Minister Jean Chretien and the leaders of Canada's provinces sign agreement to lower internal trade barriers.

26   US Congress opens hearings on Whitewater scandal involving President Clinton.

August

5   Special panel of judges removes Robert Fiske as special counsel on the Whitewater scandal and replaces him with Kenneth Starr.

September

12–   In Quebec, separatist *Parti Quebecois* wins majority in provincial government elections; Jacques Parizeau sworn in as Quebec's leader and promises referendum on independence by end of 1995 (27).

22   US Department of Defense publishes 'Nuclear Posture Review'.

October

3   US lifts ban on official contacts with Sinn Fein, and invites leader Gerry Adams to meet government officials.

3   US repeals use of 'Super 301' trade law to punish 'unfair traders'.

27   Pentagon announces termination or reduction of military operations at 27 installations overseas, mostly in Germany.

November

1   Canada announces cuts in the number of immigrants it will accept in 1995.

6   US announces it will withdraw 6,000 troops from Haiti by 1 December, and virtually all 7,800 ground troops from Kuwait by late December.

8   In mid-term Congressional elections, Republican Party wins control of both houses of Congress.

29   US House of Representatives approves legislation implementing General Agreement on Tariffs and Trade (GATT) Uruguay Round.

December

1   US Senate ratifies GATT Uruguay Round agreement.

7   Quebec's leader, Parizeau, introduces draft bill declaring Quebec a sovereign state.

9   Pentagon announces $7.7bn cut in new weapons programmes over next 6 years.

11   US announces it will cease enforcing UN arms embargo against Bosnia.

14   US agrees not to send nuclear-armed warships to New Zealand, reversing policy of refusing to confirm or deny presence of nuclear weapons on US ships.

Latin America

January

1–   In Mexico, Zapatista Army of National Liberation (*EZLN*), consisting mainly of Indians, organises rebellion in southern state of Chiapas; Mexican Army begins counter-attack (2); Zapatista rebels reject government's conditions for peace, Army seals off access to the region and bombards rebel positions (5); 4 bombs explode in Mexico City (8); President Carlos Salinas de Gortari fires Interior Minister José Patrocinio Gonzalez, and appoints Foreign Minister Manuel Camacho Solis to head a commission to bring peace to Chiapas (10); Salinas announces unilateral cease-fire and proposes general amnesty for all rebels (10); Camacho announces agenda for negotiations with rebels has been agreed (29).

6–   In Guatemala, peace talks resume between the government and left-wing rebel group; both sides agree to agenda for formal negotiations (10).

7–   Venezuelan Army helicopter lands in Colombia and soldiers allegedly kidnap 7 fishermen; Colombian President Cesar Gaviria Trujillo deploys troops to Venezuelan border (10).

31   Nicaraguan Army surrounds remaining contra rebels, killing 11.

February

6   Costa Rica's presidential elections won by José María Figueres, opposition centre-left candidate.

8–   In Nicaragua, remaining contra rebels declare cease-fire; government does likewise (9); peace talks between government and contras begin (14); peace agreement reached including full demobilisation of rebels and amnesty (25).

9–   Mexican peasant farmers take control of 4 towns in Chiapas protesting against corruption and poverty; government and Zapatista rebels open formal peace talks (22).

15–   Exiled Haitian President Jean-Bertrand Aristide rejects US-sponsored peace plan; UN Secretary-General Boutros Boutros-Ghali and mediator Dante Caputo back plan (22).

16–   Peruvian Prime Minister Alfonso Bustamante resigns; Efrain Goldenberg Schreiber, Foreign Minister, sworn in as Prime Minister (17).

28   US troops leave Colombia after tribunal rules that President Gaviria acted illegally in authorising their presence.

March

2– In Mexico, preliminary peace agreement reached between Zapatista rebels and government; presidential candidate Luis Donaldo Colosio of the ruling Institutional Revolutionary Party (PRI) assassinated (23); Ernesto Zedillo Ponce de Leon nominated PRI candidate (29).

11 In Chile, Eduardo Frei elected President.

11 At summit of Caribbean Community (CARICOM), leaders decide to seek entry into NAFTA.

20 In El Salvador, first round of general elections since end of civil war are held.

25– US Vice-President Al Gore presents revised peace plan for Haiti, but exiled Haitian President Aristide rejects it (29).

29 Guatemalan government and leftist rebels sign human-rights accord, setting up a UN human-rights mission, and agree timetable for further negotiations.

30 In Brazil, Finance Minister Fernando Henrique Cardoso resigns to register as candidate in presidential elections.

April

10 In Argentina, elections to constituent assembly, which will amend Constitution, are held resulting in setbacks for ruling party and main opposition party.

24– In second round of presidential elections in El Salvador, Armando Calderón Sol, of the ruling right-wing National Republican Alliance, elected President; Calderón holds talks with opposition FMLN leader, Schafik Jorge Handel (25).

29 US calls on UN Security Council (UNSC) to impose further sanctions on military regime in Haiti.

May

1– Cuban national assembly approves drastic austerity measures; over 100 asylum-seekers invade Belgian Embassy (28), but later leave peacefully.

6– UNSC imposes trade blockade on Haiti; Haitian legislators name Supreme Court judge Emile Jonassaint interim President (11); Dominican Republic agrees to close land border with Haiti (25).

8 In Panama, Ernesto Pérez Balladares of former ruling Democratic Revolutionary Party wins presidential election.

18 Mexico joins Organisation for Economic Cooperation and Development (OECD).

June

10– US tightens sanctions on Haiti; Haiti declares state of emergency (12); US announces most Caribbean and Latin American countries back invasion of Haiti if military regime does not step down (13).

12– In Mexico, Zapatistas reject government peace plan but rule out hostilities;

peace commission head Solis resigns citing opposition of PRI candidate Zedillo to his efforts (16).

13– Guatemalan government and rebels begin negotiations in Oslo; both sides sign agreements on resettlement of refugees (17) and establishing a commission to investigate atrocities committed during conflict (23).

13 Mexican President Salinas visits Cuba and reiterates his opposition to US trade embargo on Cuba.

14 Presidents of Colombia, Venezuela and Mexico (the Group of Three) sign free-trade agreement.

19 Colombia elects Ernesto Samper of the ruling Liberal Party President.

July

1 Brazil introduces new currency, the *real*.

11– Haiti's military regime orders international human-rights monitors to leave; UNSC condemns action (12); UNSC authorises use of all necessary means to remove military regime (31).

15 US Senate votes to make aid to Colombia conditional on its cooperation on anti-drug programmes.

24 Association of Caribbean States (ACS), an economic and political grouping with 37 members, is inaugurated.

August

1– US says Haiti's military leaders must give up power soon or be ousted; UN abandons efforts to negotiate solution to political crisis, while CARICOM leaders back US-led invasion of Haiti (30).

2– In Dominican Republic, Joaquin Balaguer proclaimed winner of presidential elections; US and Organisation of American States criticise electoral officials' decision (3); Balaguer sworn in (16).

5 Presidents of Mercosur member-countries (Argentina, Brazil, Paraguay and Uruguay) formalise agreement to create a customs union on 1 January 1995.

7 In Colombia, Ernesto Samper sworn in as President; vows to intensify war against drug cartels.

8– In Cuba, Havana's port is blocked following 4 days of clashes between police and anti-government protesters; President Fidel Castro indicates that Cubans might not be stopped from leaving if US halts its promotion of illegal departures, prompting exodus of boat people for the US (11); US reverses policy of automatic asylum for Cuban refugees, sets up detention centres at Guantanamo Bay naval base (19); US announces new policy to cut flow of US money and travellers to Cuba (20); US agrees to talks with Cuba on migration issue (27).

21– In Mexico, Ernesto Zedillo wins presidential elections; opposition leader, Cuauhtemoc Cardenas, calls for protests against alleged fraud (22).

22 Argentinian constituent assembly approves amended Constitution allowing President Carlos Saul Menem to seek re-election.

September

6 In Barbados, opposition Labour Party wins general elections.

9 Cuba and US reach agreement in which US will admit 20,000 Cuban

refugees each year and Cuba will restrict departures.

12– US persuades 17 countries to join multinational task force to invade Haiti; Clinton sends mission headed by former President Jimmy Carter to Haiti in attempt to persuade military junta to leave before US invasion (16); accord is reached allowing military leaders to stay until 15 October and granting them amnesty (18); US ground troops mount bloodless takeover of Port-au-Prince (19); US Marines deploy in northern Haiti, while forces loyal to military junta attack supporters of exiled President Aristide (20); US authorises troops to use force to halt violence by Haitian police (21); popular uprising against military regime in Cap Haitien (25); US announces lifting of most of its sanctions on Haiti and funding for reconstruction (26); explosion in Port-au-Prince during pro-democracy demonstration kills at least 5 (29); UNSC lifts sanctions against Haiti, to come into effect when Aristide resumes presidency (29); supporters of military government break up pro-Aristide march, killing at least 6 (30).

October

3– In Haiti, US troops storm headquarters of FRAPH, the nationalist paramilitary faction loyal to military government; parliament (6) and Senate (8) approve amnesty for military leaders; Lt-Gen. Raoul Cedras, head of military government, resigns (10); Emile Jonassaint, acting President, resigns (12); Cedras flies to Panama (13); US formally lifts sanctions against Haiti (14); exiled President Aristide returns to Haiti (15); UN formally lifts sanctions against Haiti (16); human-rights monitors return to Haiti (22–23); Aristide appoints Smarck Michel Prime Minister (25).

3   In Brazil, Fernando Henrique Cardoso, former Finance Minister, elected President.

10 In Chiapas, Zapatista guerrillas break off peace talks with government.

24– Cuba and US open talks to discuss migration; US federal judge bars US government from repatriating Cuban refugees held in Guantanamo Bay (25); UN General Assembly calls for lifting US trade embargo against Cuba (26).

November

8   Haiti's new Prime Minister Michel and cabinet sworn in.

17– Colombian President Ernesto Samper announces the government will hold peace talks with guerrilla groups; announces military and police appointments, in response to accusations of human-rights violations (22).

21 UN Mission for Guatemala (MINUGUA) arrives to supervise implementation of March human-rights agreement.

22 Colombia and Venezuela sign agreement to conduct joint operations against guerrilla activities on their border.

27 In Uruguay, Julio María Sanguinetti, former President and opposition party candidate, wins election.

December

1– In Mexico, Zedillo takes office as President; in Chiapas, armed Zapatista rebels announce they are occupying 38 of the state's municipalities (19);

1,000 troops and police officers move into Puerto Cate to win back rebel territory (20); government forces disperse rebels (21); government announces peso will float freely for 60 days (22); government reaches agreement with Zapatista rebels to end military operations in Chiapas and begin talks (28); Zedillo sacks Treasury Secretary and announces emergency programme to tackle severe economic crisis caused by plunge in peso (29).

8   In Panama, about 1,000 Cubans flee US-run refugee camp after riots.

9–   Three-day Summit of the Americas opens in Miami, attended by 22 presidents and 12 prime ministers, including US President Clinton, but not Cuban President Castro; Canadian, Mexican and US leaders agree to let Chile join NAFTA (11).

12   Brazilian Supreme Federal Tribunal acquits former President Fernando Collor de Mello of corruption.

21   Nicaraguan President Violeta Chamorro announces that Maj.-Gen. Joaquin Cuadra Lacayo will succeed Gen. Humberto Ortega Saavedra as Commander-in-Chief of the army.

Europe

January

1   European Economic Area (EEA) comes into force, joining the EU and five members of the European Free Trade Association (EFTA) in a free-trade area.

1   Second stage of EU's Economic and Monetary Union (EMU) begins, with establishment of European Monetary Institute in Frankfurt.

5   France announces it will send Defence Minister to some NATO meetings.

10–   NATO summit launches Partnership for Peace (PFP) initiative and Combined Joint Task Force (CJTF) concept; PFP Framework Document formally opened at NATO Headquarters in Brussels (24); Romania signs PFP document (26); as does Lithuania (27).

11–   NATO leaders reaffirm that air strikes will be carried out to prevent siege of Sarajevo and other safe areas; Croatian President Franjo Tudjman removes Mate Boban, nationalist leader of Bosnian Croats, from peace negotiations (11); Croatian Army jets and helicopters attack Muslim positions in central Bosnia (13); talks between Bosnia's 3 warring factions collapse (19); Croatia and Yugoslavia agree to normalise relations, but do not agree on sovereignty of Serb-held territory in Croatia (19); Lt-Gen. Sir Michael Rose replaces Lt-Gen. Briquemont as Commander of UN forces in Bosnia (24).

13–   Norwegian Foreign Minister Johan Jørgen Holst dies; Bjørn Tore Godal succeeds him (24).

16-   Russia's Economic Minister Yegor Gaidar resigns; as does Boris Fedorov, Finance Minister (26); President Boris Yeltsin reaffirms commitment to economic reforms (27).

26–   Belarus parliament chairman and *ex officio* head of state, Stanislau Shushkevich, ousted in vote of no-confidence; parliament elects Mechislav immediate cease-fire (18).## MarchFinland's presidential election is won by Social Democrat Martii Ahtisaari.Georgian President Eduard Shevardnadze and Russian President Yeltsin sign treaties giving Russia right to maintain bases in Georgia, deploy troops on Georgia's border with Turkey, and help train and arm Georgia's national army; Georgia and Abkhazia reiterate request for deployment of UN peacekeepers in Abkhaz region (13).Russia and Tatarstan sign treaty allowing the republic to retain its own constitution; Russian Duma grants amnesty to leaders of 1991 coup attempt and October 1993 uprising (23); Alexander Rutskoi, former Vice-President, and Ruslan Khasbulatov, former head of the parliament, are among those released from prison (26).French defence White Paper recommends closer links between France and its NATO partners.Mortar attack on Sarajevo market square kills at least 68; Bosnian Croat leader Mate Boban resigns (8); NATO orders Bosnian Serbs to withdraw artillery from 20km zone around Sarajevo within 10 days (9); President Yeltsin warns against excluding Russia from resolution of Bosnian crisis and reiterates opposition to air strikes (15); Russian special envoy, Vitali Churkin, meets leader of Bosnian Serbs Radovan Karadzic, Russian troops are deployed in exclusion zone around Sarajevo and Bosnian Serbs begin withdrawing guns from around Sarajevo (17); Bosnian Muslims and Croats agree cease-fire (23); Croatian President Tudjman backs Croat–Muslim federation (24); Croat and Muslim representatives meet in Washington (27); 4 Serb aircraft, after violating the no-fly zone, are shot down by NATO planes near Banja Luka (28).Azerbaijan and Turkey sign treaty of friendship; Azerbaijani and Armenian Defence Ministers agree to with Russian Foreign Minister Kozyrev, Bosnian Serb leader Karadzic agrees to open Tuzla airport (1); Bosnian Croat and Muslim forces begin handing in heavy weapons to UN collection points and Muslim and Croat leaders sign cease-fire maps in Croatia (7); Bosnian Army and Croatian commanders in Split sign agreement to forge a joint army (12); Bosnian government and Bosnian Croat officials agree on draft constitution (13); Bosnian and Croatian governments sign agreement forming loose confederation between Croatia and new bi-national Bosnian federation (18); Canadian UN troops find large cache of Serbian heavy weapons inside Sarajevo exclusion zone (20); UN Secretary-General approves plans to deploy Turkish peacekeepers in Bosnia (23); Serbs shell Gorazde (28–31); Croatia and Krajina Serbs sign cease-fire (30); US vetoes plan to send 8,500 extra UN peacekeepers to Bosnia (31).

6   Moldovans vote overwhelmingly not to merge with Romania.

10–   Czech Republic signs PFP document; followed by Moldova (16), Georgia (23) and Slovenia (30).

11–   Slovakia's legislature passes vote of no-confidence in Prime Minister Vladimir Meciar; he resigns (14); broad coalition government headed by Josef Moravcik sworn in (16).

15   Belarussian Supreme Soviet approves new constitution introducing presidential system.

22   IMF agrees to release $1.5bn loan to Russia.

27   In Italy's general election, the right-wing coalition of Silvio Berlusconi's *Forza Italia*, the neo-fascist National Alliance and the Northern League wins a majority of seats in the lower house and a plurality of seats in the Senate.

April

1–   Hungary applies for EU membership; as does Poland (8).

4   Georgia and Abkhazia sign rebel cease-fire agreement, and request deployment of UN peacekeeping force.

5–   Serb forces break through Goradze's line of defence; two NATO aircraft attack Serb forces around Goradze (10); cease-fire announced in Belgrade (17); Serbs continue assault (18); NATO issues ultimatum against Serbs (22); Gorazde shelled (23–24); Russian Foreign Minister Kozyrev approves use of force against Serbs who begin withdrawal (24); 'Contact Group' (US, France, Germany, the UK and Russia) formed to coordinate international effort (26); UN approves deployment of 6,500 troops to reinforce UNPROFOR (26).

5–   President Yeltsin approves plans to establish 30 permanent military bases in former Soviet republics; Moldovan parliament ratifies CIS membership (8); Russia and Belarus sign monetary union treaty (12); at CIS summit, Georgia, Armenia and Central Asian republics agree to allow Russian forces to patrol their borders (15); Russian and Latvian Presidents sign agreement for withdrawal of Russian troops from Latvia by August (30).

8–   Pro-Russian sailors in Crimea seize ship, and Ukrainian forces seize Russian-controlled base in Odessa; after second round of Ukrainian legislative elections, communists and independent deputies dominate legislature (10); talks between Ukraine and Russia on dividing Black Sea Fleet collapse (21).

13   EU Commission begins proceedings against Greece in the European Court of Justice over Greece's trade embargo on FYROM.

May

3–   Greek and Albanian Foreign Ministers fail to agree on measures to reduce tensions between their countries; Albania charges 6 ethnic Greeks with fomenting separatism (20); Greek Prime Minister Andreas Papandreou threatens to close border with Albania (30).

3–   UN Secretary-General rules out sending UN peacekeepers to Georgia; Georgia and Abkhazia agree to cease-fire and to deployment of Russian troops as CIS-sponsored peacekeepers (14).

3–   Bosnian Serbs agree to posting of 16 UN military observers in Brcko; Foreign Ministers of Russia, US and 5 EU members call for new Bosnian federation to be given 51% of Bosnian territory (13); new Bosnian federation elects Kresimir Zubak interim President (30).

4   European Parliament (EP) approves enlargement of EU to include Sweden, Austria, Finland and Norway.

4–   Azerbaijan signs PFP document, as do Sweden and Finland (9), Turkmenistan (10) and Kazakhstan (27).

9   WEU admits 9 East European states as associate members.

20–   Crimean legislature approves legislation restoring controversial constitution; Ukrainian Supreme Council legislature suspended (21); Russian and Ukrainian representatives meet in Moscow and agree to regular meetings (24).

23   Roman Herzog elected President of Germany by both houses of parliament.

26   EU-sponsored conference on the Pact for Stability in Europe opens.

29   In the second round of Hungarian general elections, the former communist Socialist Party, led by Gyula Horn, wins overall majority.

June

1–   Kyrgyzstan signs PFP document, as does Russia (22).

8–   In Geneva, warring parties in Bosnia agree on limited truce; talks between Croatian government and Krajina Serbs called off (16); Haris Silajdzic elected Prime Minister of twin parliaments of Bosnian republic and Muslim-Croat federation (23); Contact Group agrees map for division of Bosnia (29).

9–   Yeltsin signs decree creating peacekeeping force for Georgian province of Abkhazia; Russian troops deployed (24).

9–12   In EP elections held in all 12 member-states, socialists become largest single group.

12   Austrian referendum approves membership in EU by 67%.

12   Swiss voters reject government plans for UN peacekeeping operations.

14–   Ukraine and EU sign partnership and cooperation agreement; incumbent President Leonid Kravchuk leads after first round of presidential elections (28).

16   Turkish constitutional court bans pro-Kurdish Turkish Democracy Party.

21–   Latvian legislature approves law restricting Latvian citizenship; President
        Guntis Ulmanis returns bill to legislature for reconsideration (28).
24–25   At EU summit in Corfu, UK Prime Minister John Major vetoes Jean-Luc
        Dehaene as next Commission President, treaties of accession for Austria,
        Finland, Norway and Sweden are signed, and the EU and Russia sign a
        partnership and free-trade agreement.

July

5   European Court of Justice bans all exports from Cyprus not authorised by
        official Greek-Cypriot government, effectively cutting off trade between
        the EU and Turkish-Cypriot area.
6–   Contact Group announces new peace proposals for Bosnia; Bosnian Fed-
        eration approves Contact Group map (18); Bosnian Serb leaders reject plan
        (19); Bosnian Serb troops hit US and UN planes flying into Sarajevo (22).
7–   In Warsaw, US President Clinton warns that no country has a veto over
        Western integration plans; Russia participates in political deliberations at
        G-7 summit in Naples as equal partner (10).
10–   Aleksandr Lukashenka wins Belarus's first presidential election; takes
        office (20).
10–   In Ukrainian presidential elections, Leonid Kuchma defeats incumbent
        Leonid Kravchuk; is sworn in (19).
12–   German Federal Constitutional Court rules that armed military peace-
        keeping missions abroad are constitutional if parliament approves them
        first; Bundestag approves participation in the former Yugoslavia (22).
13   Uzbekistan joins PFP.
14–   Latvian government resigns; legislature passes amended citizenship law
        which still requires applicants to pass a language test (22).
15   UK and Irish prime ministers call off Anglo-Irish summit as negotiations
        over Northern Ireland stall.
15–   At emergency summit in Brussels, Jacques Santer, Prime Minister of Lux-
        embourg, chosen to succeed Jacques Delors as President of European
        Commission; EP endorses decision by slim majority (21).
18–   EU signs free-trade agreements with 3 Baltic states, while Slovenia de-
        nounces Italy's veto of an EU–Slovenian agreement; EU and Moldova sign
        partnership and cooperation agreement (26).
21   UNSC approves Russian peacekeeping force in Georgia.
28   Moldovan parliament adopts new Constitution.
29   Russian government denounces administration of Chechen republic's
        President Dzhokar Dudayev as illegitimate.

August

2–   Russian Defence Ministry announces that 14th Army, deployed in Dniestr
        region of Moldova, will be disbanded; timetable for withdrawal of 14th
        Army from Dniestr initialled by Russia and Moldova (10).
3–   Bosnian Serbs reject latest peace plan; Serbia says it is closing its borders to
        Bosnian Serbs (4); 14 NATO jets attack Bosnian Serb position near Sarajevo
        after Serb troops seize tank and heavy guns under UN control (5); US
        gives Bosnian Serbs until 15 October to accept peace plan, or it will ask UN

to lift arms embargo on Bosnian government (11); rebel Muslim enclave of Bihac overrun by Bosnian government forces (21); Bosnian Serbs reject peace plan in referendum (27–28).

6   Ukrainian President Kuchma issues decree putting himself directly in charge of the government.

9–   Russia places troops near border with Russian republic of Chechnya on higher state of alert; Chechen President Dudayev orders mobilisation to resist a Russian invasion (11).

13   NATO Secretary-General, Manfred Wörner, dies.

18   Greece strengthens measures to counter flow of illegal Albanians.

22   In the Netherlands, three-party coalition government headed by Labour Party leader, Wim Kok, sworn in.

29–   Russian troops leave Estonia, Latvia (30) and Berlin (31).

31   IRA announces complete cessation of military operations.

September

5–   Russia puts its military forces in North Caucasus military region on 'full combat alert', as Chechen President Dudayev's forces recapture town of Argun from opposition forces led by Ruslan Labazanov; Chechen opposition leader Ruslan Khasbulatov orders Dudayev to resign (29).

5   Romanian and Hungarian foreign ministers hold highest-level official talks since 1989.

8–   Serb forces from Bosnia and Croatia launch assault against Bosnian government positions in Bihac; Contact Group offers to suspend sanctions against Yugoslavia if it allows international observers to monitor its blockade of the Bosnian Serbs (8); EC foreign ministers reject lifting arms embargo against Bosnia (10–11); Bosnian and Croatian presidents meet in Zagreb and agree on key points for establishing joint federation (13); Yugoslavia agrees to allow observers to monitor its embargo (14) and the monitors begin mission (15); heavy shelling and sniper fire break out in Sarajevo (18); NATO extends air cover beyond Bosnia to part of Croatia near Bihac (21); NATO planes bomb Bosnian Serb tank near Sarajevo (22); UN votes to ease sanctions on Serbia (24).

7–   In Albania, 5 ethnic Greeks convicted of spying; Greece closes a border crossing and recalls its ambassador (8).

7–   Crimean assembly votes to strip Crimean President Yuri Meshkov of his powers; Meshkov announces dissolution of assembly (11); both actions cancelled (13); Ukrainian legislature approves constitutional amendments curbing Crimean autonomy (21); Crimean assembly confirms vote to strip Meshkov of his powers (29).

8   French, British and US troops leave Berlin.

12–   Troops from 6 NATO countries begin joint military exercises with soldiers from seven East European PFP member-states; French Defence Minister attends NATO ministerial committee (29); Belgian Foreign Minister Willy Claes appointed new Secretary-General of NATO (29).

16   UK Prime Minister Major lifts broadcasting ban on Sinn Fein, promises Northern Ireland referendum on the outcome of Anglo-Irish talks.

18   Social Democratic Labour Party returned to power in Sweden.

21–   In Denmark's general elections, government coalition loses support; Prime Minister Poul Nyrup Rasmussen forms three-party minority government and is sworn in (26).

October

1–   General elections in Slovakia end inconclusively, although Meciar's Movement for a Democratic Slovakia wins 34.96% of the vote; Meciar agrees to try to form a government (31).

2–   Turkish Foreign Minister Mumtaz Soysal condemns Greece's intention to extend its territorial waters from 6 to 12 nautical miles, as permitted under the International Law of the Sea; Greece announces postponement of the extension (30).

3–   In Bosnia, monitors report that Yugoslavia has sealed border with Bosnia and sanctions against Belgrade are relaxed; Bosnian government troops attack Serb command post on Mount Igman near Sarajevo (6); UNPROFOR troops drive Muslim forces off Mount Igman (7); UN tanks come under fire from Bosnian Serbs (26); Bosnian government troops capture more Serb-held territory east of Bihac (27); US introduces UNSC Resolution calling for lifting of arms embargo against Bosnia (28); high-level talks begin between Croatia and rump Yugoslavia on normalising relations (28); Bosnian government forces renew attacks on Serb forces in Mount Igman demilitarised zone (29); Croatian Serbs shell government positions in western Bosnia (31).

3–   In Azerbaijan, President Geidar Aliyev imposes state of emergency in capital, Baku, after Interior Ministry troops kidnap general prosecutor; more forces rebel in Gandja, Azerbaijan's second city (4); Aliyev dismisses Prime Minister Surat Guseinov for allegedly taking part in the rebellion (6).

5   Armenia signs PFP document.

9   In Austrian general election, governing coalition of Social Democrats and Austrian People's Party wins.

12–   President Yeltsin sacks Finance Minister Sergei Dubinin after rouble collapses; Central Bank governor resigns (14); Prime Minister Viktor Chernomyrdin's government survives vote of no-confidence (27).

12–   Chechen President Dudayev imposes martial law in Chechnya; opposition forces seize parts of capital, Grozny (15); government forces counter-attack (19).

13–   In Northern Ireland, loyalist paramilitaries announce cease-fire; UK Prime Minister Major announces that Sinn Fein leaders Gerry Adams and Martin McGuinness can visit UK mainland (21).

13–   Estonian parliament refuses to accept Central Bank governor, Siim Kallas, as prime minister; confirms appointment of Environment Minister Andres Tarand instead (27).

16   In referendum, Finns vote to join EU by a majority of 57%.

16   In German general elections, the ruling coalition of Chancellor Kohl's CDU, the CSU and the FDP wins with a reduced 10-seat majority.

18   Cypriot President Glafcos Clerides and President of the self-proclaimed Turkish Republic of Northern Cyprus, Rauf Denktash, begin peace talks, but make no progress.

November

3–   Bosnian government forces, supported by Bosnian Croat Army, capture Kupres in central Bosnia; UN General Assembly adopts resolution urging Security Council to lift embargo on Bosnian government (3); Bosnian Serbs announce general military call-up (4); Bosnian Serbs, supported by troops from Serb-held areas of Croatia, begin counter-attack on Bihac (9); Bosnian government troops fight Serbs on 3 fronts around Sarajevo (13); in Sarajevo, Bosnian Serb anti-tank rockets hit Bosnian government head-quarters (17); Bosnian Serb aircraft attack Bihac (18–19); NATO warplanes attack Udbina air-base in Croatia, where Serb aircraft based (21); Serbian President Slobodan Milosevic pressures Serb leaders in Croatia to halt offensive on Bihac, NATO aircraft carry out missions against Serb missile batteries in northern Bosnia (23); Bosnian Serb forces enter town of Bihac (24); detain up to 400 UN personnel (27); Croatia warns it will enter war in Bosnia if Bihac falls (29).

3   In Slovakia, Prime Minister Josef Moravcik's cabinet resigns.

6   In referendum, Albanians reject President Sali Berisha's draft Constitution.

10   Member-states of the Council of Europe approve a Convention on the Protection of National Minorities.

12– In France, Michel Roussin, Minister for Cooperation, resigns over corruption scandals, the third minister to do so since July.

13– In Swedish referendum, voters approve membership in EU with 52% majority; Finnish parliament endorses EU membership (18); Norwegian voters reject EU membership by 52% (28).

14   José Cutileiro of Portugal appointed Secretary-General of WEU.

15   German Bundestag re-elects Helmut Kohl as Chancellor.

16   Ukrainian legislature passes a law which automatically invalidates any Crimean law deemed to conflict with Ukrainian law.

17   Tensions between Greece and Turkey increase as they begin separate military exercises in disputed international waters in the Aegean Sea.

17   Irish Prime Minister Albert Reynolds and his Fianna Fail cabinet colleagues resign after Labour Party withdraws from governing coalition.

22   Italian Prime Minister Berlusconi is notified that he is under investigation for bribery.

25– In Chechnya, 40 helicopter gunships bearing Russian markings attack positions of pro-Chechen government forces near Grozny; Chechen government forces defeat assault by Russian-backed opposition forces (26); Chechen President Dudayev claims to have captured 70 Russian soldiers (27); Yeltsin orders both sides to lay down arms or face Russian military intervention (29).

26– Legislature of breakaway Abkhazia region in Georgia adopts new constitution establishing Abkhazia as a sovereign state; Georgian President Shevardnadze denounces measure (28).

December

1   NATO foreign ministers agree to study implications of enlargement, while Russian Foreign Minister Kozyrev refuses to sign two agreements detailing Russia's cooperation with NATO.

2–   Croatia signs economic agreements with Krajina Serb officials; Serbian

President Milosevic backs peace plan, but Bosnian Serb leader Karadzic refuses (4); former US President Carter meets Croatian President Tudjman, Bosnian Prime Minister Silajdzic and UN officials in Zagreb, while Velika Kladusa in north-western Bihac pocket falls to Bosnian Serb and rebel Muslim forces (18); after meeting with Bosnian Serb leadership, Carter claims they have agreed to immediate cease-fire (19); cease-fire agreement enters into force (24); rebel Muslim leader Fikret Abdic accepts UN-mediated truce in Bihac enclave (28).

2– In Chechnya, Russian parliamentary delegation tries to negotiate an end to clash with Moscow; Russia builds up tanks and armoured vehicles on Chechen border (3–4); Chechen President Dudayev and Russian Defence Minister Grachev hold last-ditch peace talks (6); Russian warplanes bomb Grozny, as Russian security council calls for Dudayev's men to disarm (7); Russian troops launch offensive (11); peace talks between Russia and Chechnya break up (12); Russia launches air strikes on Grozny (18); Russian MPs call for end to military action (23); Yeltsin announces he has ordered an end to bombing of civilians in Grozny (27); Russian troops advance on Grozny, as Russian warplanes resume bombing (28).

5– EU finance ministers approve loan of 85m ECU to Ukraine; two-day EU summit in Essen, Germany, agrees on accession strategy for 6 Central and East European countries (9–10); EU opens negotiations on association agreements with Estonia, Lithuania and Latvia (15); EU foreign ministers endorse GATT Uruguay Round agreement (19).

5–6 CSCE summit in Budapest changes its name to the Organisation for Security and Cooperation in Europe (OSCE) as of 1 January 1995.

6– Spain tightens border checks on pedestrians on border with Gibraltar; talks begin between UK and Spain over future of Gibraltar (19).

8– In Turkey, 8 Kurdish former MPs are sentenced to up to 15 years in prison for supporting the Kurdistan Workers' Party; EU refuses to ratify customs union agreement with Turkey (19).

9 UK and Sinn Fein hold first public talks.

13 In Slovakia, Vladimir Meciar sworn in as Prime Minister.

15– In Republic of Ireland, Fine Gael leader John Bruton forms coalition government and becomes prime minister; Bruton meets with UK Prime Minister Major and both pledge support for Northern Ireland peace efforts (20).

18 In Bulgaria, ex-communist Socialist Party wins general election.

21– In Italy, right-wing coalition collapses when Northern League party withdraws its support; Prime Minister Berlusconi resigns (22).

22 Ministers of 7 EU countries (Belgium, France, Germany, Luxembourg, the Netherlands, Portugal and Spain) agree to implement Schengen accord and scrap internal border controls on 26 March 1995.

24– Albanian President Berisha pardons one of the 5 ethnic Greeks convicted of treason in September, orders reduction of sentences of the others; Greek Foreign Minister Karolos Papoulias welcomes move (25).

Middle East

January

7– Jordan and PLO sign an economic cooperation agreement; Jordan's King Hussein indicates willingness to meet Israeli Prime Minister Yitzhak Rabin before a full peace is achieved (26).

15– Iraq and Jordan sign agreement on cooperation in the oil industry; head of UN inspections team to Iraq says that Iraq promised to cooperate in establishing a weapons-monitoring system (30).

16– Syrian President Hafez al-Assad meets US President Clinton in Geneva and demands Israel's total withdrawal from the Golan Heights.

18 Yemen's main political parties sign agreement to end feuding.

19– In Lebanon, Israel launches air attack on bases of the Popular Front for the Liberation of Palestine south of Beirut; a Jordanian diplomat is shot and killed in West Beirut (29).

22– PLO Chairman Yasser Arafat and Israeli Foreign Minister Shimon Peres hold inconclusive talks in Oslo; and in Davos, Switzerland (30).

February

1– Iranian President Rafsanjani escapes assassination attempt; talks open with Iraq over return of remaining Iraqi prisoners of war (18).

7– In south Lebanon, 4 Israeli soldiers ambushed by *Hizbollah* fighters and Israeli aircraft launch retaliatory strikes; 10 Christians killed at church near Beirut when two bombs explode during the service (27).

9– Israel's Foreign Minister Peres and PLO Chairman Arafat reach partial agreement on Palestinian self-rule and Israeli withdrawal from Gaza and Jericho; Israel and PLO begin new round of talks (14); talks on economic aspects of Palestinian autonomy begin (22); Israeli settler Baruch Goldstein kills at least 27 praying Palestinians in the Ibrahimi Mosque in Hebron (25); Israeli cabinet orders official inquiry into the massacre (27); bilateral negotiations between Israel and Jordan, Syria and Lebanon suspended (27).

16 PLO and Jordan agree on open border policy and economic cooperation.

20– Yemen's rival leaders, President Ali Abdullah Salih and Vice-President Ali Salem al-Baidh sign 18-point reconciliation accord; fighting erupts again between armies of the former North and South Yemens (22).

March

7– Seven Southern Lebanese Army soldiers killed in fighting in Israel's self-declared security zone in southern Lebanon and Israel responds with air attacks against *Hizbollah* targets in the area; Lebanon disbands the Lebanese Forces, a right-wing party and militia, after several of its members are arrested in connection with bomb attack on church near Beirut (23).

13– Israel outlaws fanatical Jewish groups, Kach and Kahane Chai; UNSC unanimously condemns Hebron massacre (18); informal Israeli–PLO talks re-open (21); Israel and PLO sign agreement to facilitate Israeli troop withdrawal from Gaza Strip and Jericho.

April

6–   In Israel, at least 7 people killed by a car bomb, planted by *Hamas*; Israel seals Occupied Territories in response (7); suicide *Hamas* bomber kills himself and 5 Israelis on bus in Hadera (13); Israel and PLO agree on terms for deployment of a Palestinian police force (12); Israel and PLO announce they will sign an agreement on Palestinian self-rule (28); in Paris, Israel and PLO sign agreement on economic relations (29).

14   US jets shoot down 2 UN helicopters by mistake over northern Iraq.

18–   Lebanon breaks off relations with Iraq following murder in Beirut of an Iraqi opposition figure; Samir Geagea, leader of banned Lebanese Forces, arrested over 1990 assassination of rival Dany Chamoun in 1990 (21).

27   Rival units of Yemeni Army fight battle north of San'a.

30   US Secretary of State Warren Christopher visits Syrian President Assad to discuss Israeli withdrawal from the Golan Heights.

May

4–   Israel and PLO sign deal on Palestinian self-rule in Jericho and Gaza Strip; international observers arrive in Hebron (8); first Palestinian police replace Israelis in Gaza Strip (10); Israelis complete withdrawal from Gaza Strip (18); first meeting of the partial Palestine National Authority held (26–28).

4–   Fighting breaks out between rival North and South Yemeni forces in Dhamar; North Yemeni leaders sack Prime Minister Haydar Abu Bakr al-Attas (10); South Yemen announces secession from unified Yemen (21); and forms presidential council, which elects Ali al-Baidh President (22).

17–   UNSC renews sanctions against Iraq; Saddam Hussein appoints himself Prime Minister (29).

21–   Israeli commandos abduct leader of a small Shi'i Muslim organisation in Lebanon; Lebanese government protests to UNSC (24).

June

1–   UNSC calls for immediate cease-fire and negotiations to end Yemeni civil war; UN envoy, Lakhdar Brahimi, arrives in Yemen (8); North Yemen leader President Salih orders forces to observe unilateral cease-fire (9); talks on cease-fire begin (20); both Yemens sign ceasefire (30).

2–   Israel attacks *Hizbollah* camp in southern Lebanon; further *Hizbollah* targets attacked (19–20); 3 Israeli soldiers killed in southern Lebanon (20); Israel launches retaliatory air attacks (21).

10–   Several states pledge additional $42m aid for Palestinian self-rule areas; first meeting of full Palestine National Authority held (26); Israel and PLO begin talks on extending self-rule area (28).

14   UN announces completion of programme to destroy Iraq's chemical weapons.

July

1   Arafat returns to Gaza Strip; Israeli and Palestinian leaders discuss Palestinian autonomy and create 3 joint committees (6); Arafat assumes power in Gaza Strip (12); Israel seals Gaza Strip borders after riots (18).

5–   North Yemeni leaders claim victory over secessionist southern Yemeni forces; southern rebel leaders flee southern capital of Aden (7); government forces expand control over whole of Yemen (10).

12–   Iraq launches diplomatic initiative for lifting UN sanctions; UNSC renews sanctions (18).

18–   Jordan and Israel meet to draw up peace treaty; Israeli Foreign Minister Peres holds peace talks in Jordan with Prime Minister Abdul Salamal-Majali (20); Israeli Prime Minister Rabin and King Hussein sign peace declaration (25); Syria criticises Jordan (28).

August

3–   Israeli parliament endorses peace accord with Jordan; Prime Minister Rabin meets King Hussein in first official visit to Jordan by an Israeli leader (8).

10–   Rabin meets Chairman Arafat to discuss Palestinian self-rule; Israel and PLO sign agreement to extend Palestinian authority.

September

1–   Morocco agrees to open liaison office in Tel Aviv; Syria rejects Israeli plan for staggered withdrawal from Golan Heights (8); Israel and Jordan agree to open third border crossing at Sheikh Hussein Bridge (Allenby Bridge) by end of October (12); meeting in Oslo, Israeli Foreign Minister Peres and PLO Chairman Arafat reach new 15-point accord to speed up international aid to Palestinian self-rule areas (13); Israeli Prime Minister Rabin announces plans to build new houses in a Jewish settlement in the West Bank (26); Rabin holds unannounced summit meeting with King Hussein of Jordan to draw up timetable for signing full peace treaty (29); Gulf Cooperation Council partially lifts economic boycott against Israel (30).

6–   Iraqi and Kuwaiti officials hold private talks on Kuwaitis missing since Iraq's invasion; Russia and Iraq sign bilateral trade protocol (9); Russian Foreign Ministry says trade will not be renewed until sanctions on Iraq are lifted (12).

October

2–   Tunisia agrees to first steps towards establishing diplomatic ties with Israel; Israel suspends peace talks with PLO and seals off Gaza Strip following kidnapping of Israeli soldier (11); Prime Minister Rabin and Foreign Minister Peres visit Jordan for talks with King Hussein (12); Palestinian police in Gaza Strip arrest more than 200 members o´ *Hamas* in attempt to find kidnapped Israeli soldier (13); raid by special Israeli forces to rescue kidnapped soldier in West Bank fails as soldier is killed (14); Israeli cabinet agrees to resume negotiations with PLO and to lift ban on Palestinians crossing into Israel from Gaza (16); Israel and Jordan initial peace treaty in Amman (17); Israel seals borders with West Bank and Gaza Strip following deaths of 23 people in bus bombing in Tel Aviv (19); Israel and Jordan sign formal peace agreement (26).

3–   US and UK dispatch warships to Gulf following Iraqi deployment of tens of thousands of troops near Kuwaiti border (7); UNSC reaffirms Kuwait's sovereignty, while US boosts air power in Gulf (10); Iraqi troops begin withdrawing (12); in joint Iraqi–Russian statement, Iraq offers to recognise Kuwait's sovereignty if UN promises to ease sanctions against Iraq (13); US President Clinton orders build-up of US forces in Gulf to continue after Iraqi Republican Guards halt retreat from Kuwaiti border (14); UNSC orders Iraq to withdraw all troops deployed in south (15).

19– In south Lebanon, Israeli tank attack kills several civilians; *Hizbollah* fires rockets into Israel (20–21); Israeli troops shell guerrilla targets (21).

27 US President Clinton visits Syria for talks with President Assad in attempt to overcome impasse in Israeli–Syrian negotiations.

November

1– Israel re-opens border with Gaza Strip and West Bank; Israel and PLO agree series of measure to speed expansion of Palestinian self-rule (8); Islamic suicide bomber kills 3 Israeli soldiers near Jewish settlement in Gaza Strip (11); at least 12 Palestinians killed in clashes between Palestinian police and Islamic extremists in Gaza Strip (18); in Lebanon's largest refugee camp, gun battles between supporters and opponents of PLO Chairman Arafat leave at least 8 Palestinians dead (25); in Gaza Strip more than 10,000 Palestinians attend *Hamas* rally (26); World Bank signs accord with Arafat for $58m credit to Palestinian authorities (27); US, Russia and EU promise to accelerate flow of aid to Occupied Territories (30).

6– Jordanian parliament ratifies peace treaty with Israel; Jordan and Israel establish diplomatic ties (27).

6– Iran fires at least 3 *Scud* missiles into Iraq, striking base used by exiled *mujaheddin* Khalq guerrillas; Iranian fighter-bombers strike at bases of Khalq and other Iranian opposition groups in Iraq (9).

10– Iraq formally recognises Kuwait's sovereignty and its borders; move is welcomed by UNSC (16).

December

1– Israel hands over control of taxes and health to Palestine National Authority in the self-rule areas; PLO Chairman Arafat and *Hamas* spokesman Mahmoud Zahhar announce formation of reconciliation committee (2); Israeli–Palestinian talks resume on extending Palestinian self-rule to West Bank (6); US backs Palestinian demands for Israel to move quickly towards permitting Palestinian elections (7); Arafat pledges to do whatever he can to end attacks against Israeli troops and civilians (7); Israeli Prime Minister Rabin, Foreign Minister Peres and Arafat fly to Oslo to receive their shared Nobel Peace Prize (9); Israel and Jordan open temporary embassies in each other's countries (11); Peres and Arafat resume stalled peace talks (12); Rabin says Palestinians can hold elections in West Bank and Gaza under Israeli guns or delay them for a year (14); Israeli security forces clash with Palestinian protesters as work begins on controversial expansion of Jewish settlement in West Bank (27); about 2,000 demonstrators fight Israeli soldiers during protest over expansion of Jewish settlement in West Bank (30).

5– Ali Khamenei appointed spiritual leader of Shi'i Muslims, following Ali

Arahi's death; Ali Khamenei declines leadership (14); US accuses Iran of helping Iraq to violate UN embargo imposed on its oil exports (19).

7   Yemen accuses Saudi Arabia of encroaching on its territory.

8–   In southern Lebanon, 9 pro-Israeli militiamen killed in *Hizbollah* bomb attacks; guerrillas launch attacks on Israeli occupation forces, killing 2 soldiers (19); Israeli artillery, warplanes and helicopters pound south Lebanon in retaliation (23).

13   Islamic Conference Organisation summit in Casablanca adopts code of conduct for combating Islamic militants.

19–   UN says that Iraq has misled UN inspectors investigating Iraq's biological warfare programme and has concealed radar used to track ballistic missiles; fighting erupts between main Kurdish factions in northern Iraq (23).

Asia and Australasia

January

12   Thai Prime Minister Chuan Leekpai begins 3-day visit to Cambodia.

19–   US Treasury Secretary Lloyd Bentsen visits China to discuss Chinese human-rights and trade; China and Russia sign border agreement (27).

20   In Myanmar (Burma), regime holds peace talks with Karen rebels.

21–   Upper chamber of Japanese Diet votes against political reform; committee of parliament members first fails to agree compromise formula for electoral reform package (27), and then succeeds (29).

26   US Commander-in-Chief in South Korea, General Gary Luck, requests *Patriot* missile deployment to South Korea.

27   US Senate votes to lift economic embargo on Vietnam.

31   China and Taiwan re-open 'unofficial' talks.

February

1–   US Senate calls for return of nuclear weapons in South Korea; North Korea responds by implicitly threatening to use nuclear weapons (3).

1–   US State Department report criticises China over human-rights violations; Chinese Foreign Ministry condemns report (3).

3–   Cambodian government troops capture Khmer Rouge base at Anlong Veng; Khmer Rouge recapture it (24).

3   US President Clinton lifts embargo on Vietnam.

8–   Japanese cabinet approves large economic stimulation package; Japanese Prime Minister Morihito Hosokawa meets President Clinton for unsuccessful trade talks (11).

9   Taiwanese President Lee Teng Hui begins 'private' tour of Philippines, Indonesia and Thailand.

14–   US Congressman William Richardson visits Myanmar, meets Aung San Suu Kyi, main opposition leader under house arrest; Kachin ethnic minority leaders sign cease-fire and end armed rebellion (23); Japan agrees to send aid to Myanmar (25).

20   US State Department announces resumption of high-level political contacts with New Zealand, halted since 1987.

24   Malaysian government bans contracts with British companies after corruption allegations over Pergau Dam arms-for-aid controversy appear in British newspapers (24).

24–   Hong Kong Governor Chris Patten's limited political reforms accepted by Hong Kong Legislative Council (LEGCO); Patten publishes second bill to broaden democratic franchise (25).

28   In Fiji, Sitiven Rabuka sworn in for second term as Prime Minister after general elections (18–25).

March

6–   China begins rounding up dissidents prior to official US visit by Secretary of State Warren Christopher (12–14); Japanese Prime Minister Hosokawa, visiting China, asks for assistance with North Korea (19).

8   New Zealand Prime Minister Jim Bolger proposes that New Zealand become a republic.

19   Cambodian government captures Pailin, Khmer Rouge headquarters.

19–   North Korea walks out of talks with South Korea; President Clinton agrees to deploy *Patriot* missiles in South Korea (21); South Korea places its forces on full military alert (22).

28   Singapore's Prime Minister Goh Chok Tong visits Myanmar.

April

1–   Chinese police seize leading dissident Wei Jingsheng for interrogation, before French Prime Minister Edouard Balladur visits China (7–11).

8–   Japanese Prime Minister Hosokawa resigns over corruption scandal; Foreign Minister Tsutomu Hata elected Prime Minister by the Diet (25); Social Democratic Party of Japan withdraws from governing coalition (26).

12   Taiwan suspends cultural and educational exchanges with China over its handling of a boat fire which killed 24 Taiwanese tourists in China.

18–   US *Patriot* missiles arrive in South Korea; US Defense Secretary William Perry visits Seoul to discuss allied military preparedness (19).

19   In Cambodia, Khmer Rouge recapture Pailin, their official headquarters.

May

8–   Japan's Justice Minister resigns after denying the 1937 Rape of Nanjing; US and Japan agree to resume trade negotiations (24).

21   US and Vietnam agree to open liaison offices in each other's capitals.

26–   President Clinton renews China's most-favoured nation status and abandons previous link between human rights and trade; Russian and Chinese prime ministers sign several cooperation accords (27).

27   In Cambodia, government and Khmer Rouge leaders attend conference on reconciliation and peace.

31   Non-official Asia-Pacific Conference on East Timor opens in the Philippines, straining Philippine–Indonesian relations.

June

16– Talks between Cambodian government and Khmer Rouge end with no agreement; Cambodian government orders Khmer Rouge to close its compound in Phnom Penh (17).

16 Vietnam and Russia sign friendship and cooperation treaties.

25– Japan's Prime Minister Hata resigns after coalition collapses; socialist leader Tomiichi Murayama elected Prime Minister after Liberal Democratic Party and Social Democratic Party form coalition (29).

30– In Hong Kong, LEGCO approves political reforms; Britain and China settle dispute over transfer of Hong Kong military sites (31).

July

2– Cambodian government foils attempted *coup d'état*; Khmer Rouge proclaims provisional government in part of northern Cambodia (11).

5– Taiwanese official document reiterates policy of seeking reunification with China on Taiwan's own terms; Taiwan's national assembly approves constitutional reforms (29).

8 North Korean President Kim Il Sung dies.

12 Russian and Chinese defence ministers sign agreement to avert military accidents.

14 In East Timor, Indonesian soldiers attack student demonstration in Dili.

25 Asia-Pacific foreign ministers, including that of the US, hold first formal meeting of the ASEAN Regional Forum to discuss regional security problems.

August

4– Chinese delegation begins visit to Taiwan; senior officials reach agreement to repatriate airline hijackers and to resolve fishing dispute (8).

22 Japanese Prime Minister Murayama begins tour of Philippines, Vietnam, Malaysia and Singapore apologising in each country for Japan's conduct during the Second World War.

27 US Commerce Secretary Ronald Brown visits China and agrees to take human rights off the public agenda.

31 China issues law under which Hong Kong's political reforms will be abolished as soon as China regains control on 1 January 1997.

September

2– Chinese President Jiang Zemin begins first visit to Russia by a Chinese president since 1957; Chinese and Russian foreign ministers sign agreement demarcating part of their mutual border (3); Russian border troops shoot dead 2 crewmen on Chinese fishing boat in southern Kuril Islands (11).

2– In Papua New Guinea, Prime Minister Sir Julius Chan holds peace talks with Bougainville secessionist rebels; cease-fire is agreed (8).

7– US upgrades relations with Taiwan; China registers 'strong protest' to US over its decision (10); UN committee rejects Taiwan's request for its membership to be placed on General Assembly agenda (21).

7   Malaysia lifts ban on UK companies seeking government contracts.

15–   UK Foreign Secretary Douglas Hurd visits Hong Kong; Democratic Party emerges as strongest party in first fully democratic local elections there (18).

20   In Myanmar, military junta meets detained pro-democracy leader Aung San Suu Kyi for first talks since start of her house arrest 5 years ago.

27   East Timorese guerrillas clash with Indonesian troops.

October

1   Japanese and US trade negotiators reach agreements in several areas.

3–   China's Vice-Premier and Foreign Minister visit US and sign accord agreeing not to sell medium-range missiles abroad (4); US Defense Secretary arrives in China to discuss defence cooperation (16); US and China sign agreement on conversion of Chinese defence industries (17).

6   Indonesian Foreign Minister holds talks in New York with a leader of the East Timorese resistance.

12–   Vietnam says it will join ASEAN next year; Vietnam and China exchange complaints about behaviour in the Spratly Islands (17).

16   Papua New Guinea's government abandons talks with Bougainville rebel movement.

25   Cambodian government troops overrun key Khmer Rouge base in search of 3 Western hostages.

29–   Myanmar military leaders meet Aung San Suu Kyi; high-level US delegation arrives in Myanmar to discuss US–Myanmar relations (31).

November

2–   Lower house of Japan's Diet approves electoral reform bill, as does upper house (21); Japan postpones plans to send mission to North Korea to discuss improving bilateral relations (29).

3–   Indonesian President Suharto offers to meet exiled Timorese leaders; East Timorese students invade and occupy US Embassy in Jakarta demanding US help to release guerrilla leader José Xanana Gusmao (12); President Clinton visits Indonesia for APEC meeting, and raises issue of human rights with Suharto (16); students end sit-in at US Embassy, leave Indonesia for exile in Portugal (24).

4   Hong Kong and China conclude agreement on financing Hong Kong's new airport.

6–   China says it has released 4 political prisoners and 4 Tibetans; Canada and China sign deals on nuclear power and friendly aid (7); US Defense Intelligence Agency Director makes low-profile visit to China (7); Chinese President Jiang Zemin visits Vietnam, agrees to try to settle border dispute and refrain from use or threat of force (22).

7   South Korea lifts ban on direct trade and investment in North Korea.

8–   Japan resumes aid to Myanmar; General Bo Mya, leader of rebel Karen group, says in Bankok he is ready to discuss cease-fire (30).

14–   Taiwan accidentally shells village in south-east China during military exercises; Taiwan apologises to China (15); talks are held on technical issues (22–27).

15   18 Asia-Pacific countries, members of APEC, begin summit meeting in Indonesia and agree to work towards creating world's largest trade area.

15   In Cambodia, Khmer Rouge claim they executed 3 Western hostages in October, and 2 others who had disappeared in April.

December

5–   US Transportation Secretary Federico Pena visits Taiwan's Foreign Ministry; China cancels planned visit by Pena in protest (13).

7   Cambodian government troops capture Khmer Rouge stronghold, Phnom Kulen, in north-west.

15–   US officials suspend talks with China over copyright infringement; Chinese Prime Minister Li Peng arrives for 3-day visit to Myanmar (26).

17–   US Army helicopter shot down when it strays into North Korean air space; North Korea returns body of one pilot, surviving pilot is not yet released (22); US sends senior diplomat to North Korea in further attempt to secure release of pilot (26); US pilot freed (30).

South and Central Asia

January

1–   In Afghanistan, fighting breaks out between forces of President Burhanuddin Rabbani and Prime Minister Gulbuddin Hekmatyar and his ally, General Abdul Rashid Dostum; warring factions agree to temporary cease-fire for diplomats to leave Kabul (8).

2–   India and Pakistan hold high-level direct talks on Kashmir with no progress; India submits further proposals to ease border tensions, including an agreement not to use nuclear weapons first (24).

10   Kazakhstan and Uzbekistan agree to create economic union.

15   In referendum, Turkmenistan's voters support extension of President Saparmurad Niyazov's term.

30   In Kyrgyzstan, voters approve President Askar Akayev's referendum on economic reforms.

February

2   China and India begin 3 days of talks on border troop reductions.

13   Kazakhstan President Nursultan Nazarbayev begins visit to US and obtains increase in US aid to help dismantle nuclear weapons.

14–   Afghan factions agree to truce to allow food to reach Kabul; cease-fire breaks down (15); Pakistan's Foreign Minister begins efforts for peace (18); demonstrators attack Pakistani Embassy in Kabul protesting at Pakistan's alleged support for Hekmatyar (23); Pakistan cancels all passports for Afghan refugees (25).

21   Pakistan commandos storm Afghan Embassy in Islamabad killing 3 Afghan gunmen and freeing 3 schoolboys held hostage for 2 days.

March

6–   Afghan Prime Minister Hekmatyar allows UN food convoy to enter Kabul; 5-day cease-fire begins as UN-sponsored peace talks open in Jalalabad (30).
7–   In Kazakhstan, parliamentary elections are held with a clear victory for supporters of President Nazarbayev; he visits Russia and signs several cooperation agreements, including one on dismantling nuclear weapons in Kazakhstan (29).
11–   Tajik Deputy Prime Minister and chief government negotiator with rebels assassinated; talks between government and rebels postponed (11); foreign ministers of Russia, Tajikistan, Kazakhstan, Kyrgyzstan and Uzbekistan request that CIS peacekeepers in Tajikistan be given UN status (15).
20–   Pakistan shuts Bombay consulate as tensions rise over Indian human-rights violations in Kashmir; India closes consulate in Karachi (22).

April

11–   Afghan cease-fire collapses in Kabul; second round of UN peace talks held (16).
14–   Tajik opposition rebels clash with Russian troops; Tajik government and opposition agree to cooperate on return of refugees and to seek national reconciliation (19).
18–   Chinese Prime Minister Li Peng visits Uzbekistan, Turkmenistan, Kyrgyzstan and Kazakhstan; says China wants to boost regional trade (19).

May

29   Kyrgyzstan joins economic union of Kazakhstan and Uzbekistan.

June

24   Sri Lankan President D. B. Wijetunge dissolves parliament and announces new elections.
26   Afghan rebels loyal to President Rabbani capture Kabul.
28   Talks in Tehran between Tajik government and opposition end without much progress.
29   Indian Prime Minister Narasimha Rao begins visit to Russia, signs military cooperation agreements.

July

1   New currency, the *som*, introduced in Uzbekistan.
6–   UN announces that the parties to the conflict in Afghanistan do not favour a UN peacekeeping force; Organisation of the Islamic Conference (OIC) launches peace initiative (11).
9   Uzbekistan, Kyrgyzstan and Kazakhstan agree to strengthen economic and military cooperation.
10–   Nepalese Prime Minister Girya Prasad Koirala resigns following vote of no-confidence; King Birendra dissolves parliament (11).

15   Kyrgyzstan's President Askar Akayev orders recognition of Russian as an official language.

19   Chinese Foreign Minister Qian Qichen completes visit to India with no progress made on border talks.

20   Tajik legislature approves draft constitution.

August

8   Indian government withdraws troops from holiest Muslim shrine in Kashmir, but extends emergency powers for a further 6 months.

16–   In Sri Lanka, Chandrika Kumaratunga's People's Alliance defeats ruling United National Party in elections; coalition government sworn in with Kumaratunga as Prime Minister (19); government eases economic blockade of region controlled by Tamil Tigers (31).

18   On Tajik–Afghan border, 10 Tajik Army troops and 7 Russian border guards killed in separate clashes with Tajik rebels.

September

5–   In Sri Lanka, new government lifts nationwide state of emergency; Tamil Tigers accept government's offer of peace talks (7); fighting flares up again (8–16); Tamil suicide squads ram and sink navy patrol boat, killing 25 Sri Lankan sailors (20); ruling People's Alliance selects Prime Minister Kumaratunga as its presidential candidate (23).

5   Kyrgyzstan's government resigns, and President Akayev orders new parliamentary elections by end of 1994.

12–   In Afghanistan, President Rabbani's forces launch air strikes against opposition positions outside Kabul; the forces lose town of Khenjan to forces loyal to Prime Minister Hekmatyar (16); UN peace envoy starts talks with several Afghan factions (29).

13   India rejects UN offer of mediation in Kashmir dispute with Pakistan.

17–   Tajik government and opposition leaders agree to temporary cease-fire at UN-sponsored talks in Tehran; Russian border guards kill 2 Tajik rebels in gun battle close to Afghan border (23).

October

11–   In Pakistan, 13 killed in unrest during national strike; 2 days of religious and ethnic violence in Karachi leave 50 dead (18).

11–   Kazakh government resigns over slow pace of economic reforms; President Nazarbayev names Deputy Prime Minister Kazhageldin Akezhan Magzhan Ulu new Prime Minister (12).

12–   In Sri Lanka, Tamil rebels and government soldiers clash; government and Tamil negotiators start 2 days of talks in Jaffna (13); leader of United National Party, Gamini Dissanayake, and 53 others assassinated in bomb blast at presidential election rally in Colombo (24); government declares state of emergency and suspends peace talks with Tamil rebels (24); United National Party chooses Dissanayake's widow, Srima Dissanayake, as presidential candidate (25).

14   Indian government releases from prison one of Kashmir's most popular

political leaders, Shadir Shah.

18   Presidents of Azerbaijan, Kazakhstan, Kyrgyzstan, Turkmenistan, Turkey
      and Uzbekistan meet in Ankara to discuss economic and cultural coopera-
      tion.

24–   In Afghanistan, cease-fire is mediated by Iran, but opponents of President
      Rabbani fire rockets against Kabul (24 and 31).

November

1–   Indian police free 3 British tourists held hostage by Kashmiri militants;
      Prime Minister Narasimha Rao assumes personal charge over Jammu and
      Kashmir Affairs department (31).

1–   Tajikistan's government and opposition agree on joint commission to
      oversee cease-fire; acting head of state Imamali Rakhmonov wins presi-
      dential elections (6).

6–   Five Afghan factions loyal to President Rabbani accept UN peace plan
      providing for immediate cease-fire; Afghan peace talks sponsored by the
      OIC open in Tehran (29).

8–   Sri Lankan Navy kills 7 Tamil rebels in sea battle; Prime Minister
      Kumaratunga wins presidential election (9); Tamil Tigers unilaterally an-
      nounce a week-long cease-fire (12); President Kumaratunga appoints her
      mother, Sirimavo R. D. Bandaranaike, Prime Minister (14).

8–   In Pakistan, militants release 60 hostages abducted to press their demand
      for Islamic law in northern areas; 17 killed in gun battles between rival
      political factions in Karachi (11).

15–   In Nepalese parliamentary elections, Communist Party wins most seats
      but not a working majority; Man Mohan Adhikary elected parliamentary
      leader (23); King Birendra names Adhikary Prime Minister (29).

December

2–   Tajik President Rakhmonov names Jamshed Karimov Prime Minister in
      cabinet re-shuffle; Rakhmonov orders nationwide disarmament (3); UNSC
      establishes UN Mission of Observers in Tajikistan to oversee cease-fire
      (16).

11   Elections held in Turkmenistan, but all candidates run unopposed and as
      members of the ruling Turkmen Democratic Party.

12–   In Sri Lanka, Tamil rebels accept government proposals for cease-fire;
      Tamil rebels kill 3 soldiers, but rebel leaders and government officials
      continue discussions for next round of peace talks (19).

15–   In Karachi, gunmen kill 10 and fire shots in front of Prime Minister
      Benazir Bhutto's home; strikes, called by a militant Sunni Muslim group
      and transport union, paralyse Karachi and 8 are killed (18); Pakistan
      orders India to close its consulate in Karachi, saying it is 'sponsoring
      terrorism' (26).

22–   Indian Prime Minister Rao, sacks 3 cabinet ministers accused of corrup-
      tion; Muslim separatists kill Hindu politician in Jammu and Kashmir
      state, sparking Hindu mob violence (27).

28–   In Bangladesh, all opposition members of parliament resign in attempt to
      oust government; opposition parties reject Prime Minister's offer to resign
      one month before next election (30).

Africa

January

3–   Two UN agencies evacuate Mogadishu, Somalia, after wave of attacks on humanitarian groups; last 8 detainees held by UN forces freed (18).

3–   Nigerian troops occupy 2 Cameroonian islands; Nigerian Foreign Minister, Baba Gana Kingibe, meets Cameroon's President Pau Biya and denies border violation (6).

5   Juvenal Habyarimana sworn in as Rwanda's interim President.

5   Togo imposes curfew on capital, Lome, after 40 die during alleged attempt to assassinate President Gnassingbe Eyadema.

6–   Peace talks between Angolan government and rebel UNITA begin, although fighting continues; they agree to create a neutral police force (30).

10   Nigerian leader General Sanni Abachi announces abandonment of market reforms.

10– Rival Sudanese rebel factions agree to cease-fire; large build-up of government forces in southern Sudan reported (24).

13   In elections in Burundi, Cyprien Ntaryamira elected President.

14– UNSC expands mandate of UN Observer Mission in South Africa to include observation of elections; Pan-Africanist Congress (PAC) suspends armed activities (16); talks between ANC, government and right-wing Freedom Alliance break up over Alliance's refusal to play a part in transition (31).

14– In Lesotho, rival army factions fight; heavy fighting in capital, Maseru (23); Presidents Mugabe of Zimbabwe and de Klerk of South Africa and ANC leader Nelson Mandela agree to set up task force for Lesotho (26).

20   Cameroon accuses Chadian rebels of killing 9 security force troops in north Cameroon.

25– In Algeria, a state-sponsored national reconciliation conference is convened but cannot agree on a new head of state and is not attended by main opposition parties; Brig.-Gen. Liamine Zeroual appointed as new head of state by High Security Council (30).

30   In Congo, opposition and government agree to a cease-fire and to disarm their militias.

February

1   Lesotho's feuding army factions in Maseru return to barracks.

2–   Mali's Prime Minister Abdoulaye Sekou Sow resigns; Ibrahim Boubakar Keita appointed instead (4).

2–   In Egypt, police kill 7 suspected Islamic militants in raid, while main Islamic militant group warns foreign investors to leave the country; luxury trains bombed (19 and 23).

2–   South African President de Klerk sets first democratic election for 26–28 April; 19 parties register, but none of the parties that form the Freedom Alliance (12).

3   International Court of Justice rejects Libya's claim to 114,000km^2 of Chad's territory, the Aouzou strip.

4–   UNSC approves new mandate for UN Operation in Somalia (UNOSOM II), emphasising peacemaking and reconstruction; heavy fighting erupts in Kismayu (11).

7–   In Togo, first round of multi-party legislative elections held; second round leaves opposition parties and ruling Rally of the Togolese people tied (20).

7–   Algerian security forces kill head of Armed Islamic Group (GIA).

8   Sudan's military government begins offensive against rebel Sudan People's Liberation Army (SPLA) in south.

17   Angolan government and UNITA sign document of 5 principles of national reconciliation.

22   In Rwanda, swearing-in of new government indefinitely postponed after Felicien Gatabazi, Minister of Works, assassinated.

28   Walvis Bay transferred to Namibia by South Africa.

March

2   French military mission arrives in Cameroon at government's invitation because of dispute with Nigeria over Bakassi peninsula.

6–   In Burundi, Tutsi-dominated army accused of killing 200 people, mainly Hutus; heavy fighting erupts between government troops and Hutus (21).

7   In Liberia, transition period begins with inauguration of 5-member Council of State and legislative assembly.

8–   Violence flares in South African bantustan of Bophutatswana after its president, Lucas Mangope, bans involvement in South African elections; police and soldiers revolt against Bophutatswana government (10); South Africa's Transitional Executive Council takes control of Bophutatswana (12); Inkatha officials say the party will not participate in elections unless their constitutional demands are met (15); in Johannesburg, violence erupts at an Inkatha anti-election demonstration and 31 die (28); de Klerk imposes state of emergency in Kwazulu-Natal (31).

10–   In Algeria, 1,000 Islamist prisoners escape from top-security prison; dialogue between President Zeroual and several political parties, including Islamic Salvation Front (FIS), begins (19).

11–   In Somalia, UN-sponsored talks begin; Somali factions sign agreement to end civil war (24); last major US combat unit leaves (25).

12–   UN Secretary-General advances new peace proposals for Western Sahara; rejected by the pro-independence Polisario Front (25).

14–   Togo legislative elections results published giving two opposition parties a majority; President Gnassingbe Eyadema refuses to endorse coalition's choice of prime minister (28).

20   In Tunisian elections, President Zine al-Abidine Ben Ali and his ruling party re-elected.

April

6–   Presidents of Rwanda and Burundi, Juvenal Habyarimana and Cyprien Ntaryamira, killed in air crash; mass fighting breaks out in Kigali with Rwandan Prime Minister, Agathe Uwilingiyimana, and 10 Belgian UN soldiers killed (7); Belgian and French troops arrive in Kigali to rescue Westerners (9); rebels of Rwandan Patriotic Front (RPF) close in on Kigali (11); Rwandan government flees Kigali (12); Belgium pulls out of UN contingent (14); UNSC reduces Assistance Mission to Rwanda (22).

11–   Mozambique President Joaquim Chissano announces multiparty elections for October; government and Renamo agree to create new Mozambique Defence Armed Forces (16).

11   Algerian President Zeroual replaces Prime Minister Redha Malek with Mokdad Sifi after Malek and cabinet resign.

12–   Lord Carrington and Henry Kissinger arrive in South Africa to mediate between ANC and Inkatha; Mangosuthu Buthelezi, President of Inkatha Freedom Party, calls off election boycott (19); bomb explodes in Johannesburg near ANC headquarters (23); white minority parliament votes itself out of power (25); first non-racial elections held (26–29); heads of international observer missions declare elections were free (30).

15   Somali factions postpone meeting to prepare national reconciliation conference.

18   Organisation of African Unity (OAU) mission arrives in Cameroon to try to settle conflict with Nigeria over Bakassi peninsula.

19–   In Burundi, clashes break out between Tutsi-dominated armed forces and Hutus; coup attempt by Tutsi paratroopers fails (25).

20   Angolan government and UNITA reach agreements on second round of presidential elections.

28   Polisario accepts UN resolution on holding a referendum on future of Western Sahara.

May

6–   South African election results announced, giving ANC huge victory; newly elected parliament chooses ANC leader Mandela as President of South Africa (9); Mandela sworn in (10).

9–   UN describes mass killing of mainly Tutsis in Rwanda as pre-planned assassinations; UNSC authorises 5,500 additional peacekeeping troops for Rwanda (17); RPF takes Kigali airport (22); UN envoy Iqbal Riza meets RPF chief General Paul Kagame (26); Rwandan government army chiefs and rebels discuss cease-fire (30).

16–   Malawi Congress Party accepts new Constitution; parliamentary and presidential elections won by opposition candidate, Bakili Muluzi, and opposition United Democratic Front (17); Muluzi sworn in (21).

16   Liberian factions agree on composition of national transitional government which meets for the first time.

23   In Nigeria, first round of elections for delegates to constitutional conference boycotted by pro-democracy groups.

26–   In Cuito, Angola, fighting erupts between government and UNITA; government offers UNITA 4 ministerial portfolios (30).

27   Somali factions again postpone national reconciliation conference.

31   Libyan troops complete withdrawal from Aouzou strip in Chad.

June

1   South Africa rejoins Commonwealth.

2–   Rwandan Army and RPF fail to agree cease-fire at talks; RPF forces seize Gitarama (13); warring sides agree to cease-fire, but then truce talks break down (15); UNSC approves France's proposed humanitarian intervention (22); French troops enter Rwanda (23).

3–   IMF suspends Zaire's membership for failing to repay debts; President Mobuto Sese Seko appoints Kengo Wa Dondo Prime Minister (17).

4   Libya and Chad sign friendship treaty.

5   In Ethiopia, elections to a constituent assembly are held.

11–   Nigerian opposition leader, Moshood Abiola, declares himself President and head of government; Abiola disappears, and police begin hunt for him (12); Nigerian police arrest Abiola (23).

12   Djibouti government and rebel Front for the Restoration of Unity and Democracy (FRUD) agree to end civil war.

12–   Sudanese government forces recapture Kajo Kaji; Sadiq al-Mahdi, leader of banned opposition party, is arrested and accused of conspiring against the government (20).

19–   Leaders of 19 Somali clans agree to end fighting in southern Somalia; heavy fighting in Mogadishu erupts (24–26).

28   Angolan government and UNITA sign document of national reconciliation, although fighting continues.

July

4–   In Rwanda, RPF rebels seize Kigali and Butare; France begins to equip force of 500 soldiers from 5 African nations (Senegal, Congo, Guinea-Bissau, Chad, Niger) to replace French troops in Rwanda (15); RPF takes last government stronghold, Gisenyi, and claims victory (18); RPF announces formation of broad coalition government, headed by Hutu, Faustin Twagiramungu (19); US helps French soldiers supply clean water to refugees as cholera epidemic spreads (21); US re-opens Kigali airport to deliver international humanitarian aid (24).

4–   French President François Mitterrand begins visit to South Africa; South African Finance Minister Derek Keys resigns, and is replaced by Christo Liesenberg (5); ANC and Inkatha Freedom Party meet and call for cease-fire (14).

5   In Ethiopia, official election results give ruling Ethiopian Peoples' Revolutionary Democratic Front a landslide victory.

6–   In Nigeria, opposition leader Moshood Abiola charged with treason; he is refused bail (14); police disrupt pro-democracy march led by Noble laureate Wole Soyinka and oil workers vow to continue 'indefinitely' strike action until government meets its demands (24); Abiola goes on trial, while police fire on pro-democracy protesters, killing 3 (28).

7   South African President Mandela hosts meeting of presidents of Angola, Mozambique and Zaire in attempt to end Angola's civil war.

9–   In Mozambique, Renamo guerrillas block national highway; South African President Mandela visits Mozambique, his first state visit (20–22); Renamo urges government to demobilise all soldiers at assembly points and postpone plans for unified army, following widespread mutiny by government troops and former guerrillas (28).

11–   In Algeria, Muslim extremist gunmen kill 11 people, including 7 foreigners; Italy announces reduction of diplomatic staff in Algiers (12); Yemeni and Omani ambassadors kidnapped by Islamic militants (15) and released (23).

22–   In Gambia, soldiers rampage through Banjul demanding back pay, and oust President Sir Dawda Kairaba Jawara (23); Jawara flees to Senegal aboard US warship (24); new President Lt Yayah Jammehname's government drawn equally from soldiers and civilians (26).

August

3– In Algeria, 5 French nationals killed in attack on French Embassy; France detains 16 Algerian Muslim fundamentalists (5); President Zeroual meets leaders of 5 legal opposition parties (21 and 24); Morocco imposes entry visas for Algerians (26) and Algeria reciprocates (28).

3– In Nigeria, general strike begins in protest against detention of Abiola; Nigeria Labour Congress suspends general strike to allow negotiations with government, but oil workers continue strike (4); government dissolves leadership of oil unions and Nigerian Labour Congress (17); arrests 25 opposition leaders (19).

8 In Burundi, strikes and clashes break out in Bujumbura after leader of radical party is arrested.

9 France and Belgium resume cooperation with Zaire as a result of it's support for international humanitarian mission in Rwanda.

9– Angolan government signs procedural agreement with UNITA; UNSC defers further sanctions on UNITA (12).

11 Chad government and southern rebels sign peace agreement, including immediate cease-fire.

16 New Mozambique Defence Armed Forces, made up of government and rebel forces, inaugurated.

17– In Lesotho, King Letsie III dismisses Prime Minister Ntsu Mokhehle, dissolves parliament and suspends parts of the Constitution, sparking large protest outside royal palace; general strike in support of Mokhehle (22); Zimbabwe, South Africa and Botswana demand Mokhehle's reinstatement (23).

20– Zaire closes its border with Rwanda at Bukavu after thousands of Rwandan Hutus flee to Zaire; France withdraws its last troops from Rwanda, handing control over to the UN Assistance Mission for Rwanda (21).

22 In Somalia, 6 UN peacekeepers killed in an ambush, while President Mohammed Ibrahim Egal of Somaliland expels UNOSOM from his self-proclaimed independent state.

28 In Western Sahara, UN voter identification programme for planned referendum on self-determination begins.

29 South Africa formally enters the Southern African Development Community.

September

4– In Nigeria, labour leaders call off strike by oil workers; government announces measures to extend period it may detain suspects without trial and bans three newspapers (6); Nigerian soldiers attack Cameroonian troops in Bakassi peninsula (8); opposition leader Abiola's trial suspended pending a suit challenging court's jurisdiction (21); government sets up an enlarged military-dominated ruling council (27).

5–   In Rwanda, RPF soldiers to be deployed in south-west under UN control; head of UN team investigating massacre in Rwanda resigns as UN member-governments fail to donate funds to operation (11); UNHCR suspends policy of repatriating refugees, following claims of massacres by RPF (23); UN aid agencies pull all international staff out of Rwandan refugee camp in eastern Zaire after bandits take control of camp (30).

5–   Angolan government and UNITA agree to new mandate for UN Angola Verification Mission; UNSC says will not strengthen sanctions against UNITA (9).

6–   In Algeria, jailed leader of FIS accepts conditions for peace talks set down by President Zeroual; government releases three FIS leaders (13); talks held between government and several political leaders, but FIS boycotts them (20); security forces shoot dead leader of militant Armed Islamic Group (26).

7   Talks between Sudanese government and factions of rebel Sudan People's Liberation army end in deadlock.

7–   Leaders of Liberia's 3 main factions meet to discuss peace settlement; Charles Taylor, leader of National Patriotic Front, ousted by members of his faction (7); 3 factions sign new peace treaty providing for cease-fire and elections within a year (12); fighting erupts again in Monrovia (15); Liberian civilians capture leader of failed coup attempt, Charles Julue, and hand him over to African peacekeeping troops (16).

10–   In Burundi, power-sharing agreement signed by 9 parties; government troops and gunmen from self-declared Popular Democratic Hutu Army clash (13 and 14); convention of parties appoints interim president, Sylvestre Ntibantunganya, as President (30).

14   In Lesotho, ousted Prime Minister Mokhehle reinstated.

15   Last US troops leave Somalia.

15–   US allocates $1,000,000 for Renamo's election campaign finances; Renamo leader lifts threat to boycott next month's elections (21).

17   In Egypt, convoy of UN aid workers attacked and 4 policemen killed.

20   In South Africa, Zulu King Goodwill Zwelithini cuts links with Chief Mangosuthu Buthelezi.

28   In Niger, President Mahamane Ousmane appoints Souley Abdoulaye new Prime Minister.

October

1–   In Somalia, fighting breaks out between rival groups in Mogadishu.

6–   In speech to UN General Assembly, Rwandan President Pasteur Bizimungu denies RPF is carrying out revenge killings of returning Hutu refugees; Belgium reinstates embassy in Kigali (26).

9–   In Niger, government and representatives of Tuareg rebels sign agreement to end conflict; President Mahamane Ousmane dissolves parliament, following motion of no-confidence in new Prime Minister, Abdoulaye Souley (16).

12   Moroccan Prime Minister Abdellatif Filali says UN referendum on disputed Western Sahara could be held in February or March.

12–   EU cuts aid to the military regime in the Gambia; US does likewise (28).

12–   Chad government and rebel Chadian National Front sign peace agreement; rebel leader challenges agreement (16).

13	South African President Mandela appeals to mutinous troops from ANC military wing and Azanian People's Liberation Army to return to base.
14	In Egypt, Nobel laureate Naguib Mahfouz stabbed by a suspected Islamic militant.
21	UNSC reduces strength of the UN Observer Mission in Liberia, and ECOWAS reduces forces deployed with its Cease-fire Monitoring Group.
21	In Nigeria, Federal High Court declares that detention of opposition leader Abiola is illegal.
27–	In Mozambique, voting begins, but Renamo pulls out of election; after talks with UN officials and Zimbabwean President Robert Mugabe, Renamo calls off election boycott and voting is extended (28); voting ends (29).
29–	In Algeria, government announces impasse in contacts with Muslim fundamentalists; President Zeroual announces that presidential election will be held by end of 1995 (31).
31	Angolan government and UNITA initial agreement to end civil war.

November

1–	In Algeria, Islamic radicals dismiss President Zeroual's pledge to hold elections before end of 1995, as FIS President Abassi Madani is re-imprisoned; various Algerian parties, including FIS but not the government, attend talks in Rome (22 and 23).
3–	UN called on to intervene to restore security for Rwandan refugees and aid workers threatened by Hutu gangs in camps in Zaire, Burundi and Tanzania; UNSC votes to set up International Criminal Tribunal for Rwanda to try those responsible for genocide (8).
4	In Nigeria, Federal Court of Appeal grants unconditional bail to opposition leader Abiola.
4–	UNSC decides to withdraw all peacekeeping troops in Somalia by 31 March 1995; secessionist rebels in northern Somalia attack regional capital, Hargeisa (17).
7	South African government dismisses over 2,000 insubordinate former members of ANC military wing from the new SADF.
10–	Angolan government captures Huambo from UNITA; government and UNITA sign truce (15); they sign peace treaty, but UNITA leader Savimbi is not present (20); Savimbi endorses agreement and agrees to meet President José Eduardo dos Santos (23).
11	In Gambia, military government foils coup attempt by junior army officers.
11	Defence ministers of 11-member Southern African Development Community agree to form regional rapid deployment force.
19	In Mozambique, President Chissano and his Frelimo party win elections; rebel leader Dhlakama accepts result.
25–	In Sierra Leone, military government calls for talks with rebel Revolutionary United Front to end fighting; rebels reject offer (28).

December

2–	UN report states that genocide in Rwanda cost 500,000 lives and that hardline supporters of former President Habyarimana and his govern-

ment were behind it; UN troops sweep through Rwanda's biggest Hutu refugee camp in dawn raid to arrest extremist militiamen and seize weapons (14); UN war-crimes prosecutor, Richard Goldstone, arrives in Rwanda (19).

2– Burundi's main opposition party, Unity for National Progress, withdraws deputies from national assembly in protest at election of new parliamentary speaker, Jean Minani; Burundi government orders curfew in Bujumbura after fighting between Hutus and Tutsis leaves at least 30 dead (22).

5– In Nigeria, Federal Court of Appeal suspends court judgment releasing Abiola; conference framing new Constitution decides military should relinquish power at start of 1996 (7).

6 Eritrea breaks off diplomatic relations with Sudan, accusing its military government of seeking to destabilise Eritrea's government.

6– In Somalia, UN helicopter gunships and tanks fire on Somali militiamen attacking Bangladeshi peacekeepers as they withdraw from a base near Mogadishu; Indian warships steam into Kismayu to cover withdrawal of last 850 Indian peacekeepers (7); fighting resumes between rival militias in Mogadishu (19).

7–8 Presidential and parliamentary elections in Namibia result in large victory for ruling South-west Africa People's Organisation (SWAPO).

22 In Liberia, 6 major warring parties agree on a new peace.

24– At Algiers airport, 4 armed men hijack aeroplane bound for Paris, demand release of imprisoned FIS leaders; aeroplane flies to Marseille, France, where French anti-terrorist police storm plane, killing the 4 terrorists (26); France suspends air and sea passenger links with Algeria, pending new security measures (27); 4 foreign priests murdered in Algeria by Islamic terrorists (27).

International Organisations and Arms Control

January

14 Ukraine, Russia and the US sign agreement under which Ukraine will transfer all nuclear warheads on its territory to Russia for dismantlement in exchange for financial compensation and security guarantees.

14 US and Russia announce that their weapons are no longer aimed at each other's countries.

15 Indian government spokesman says India will reject extension of the NPT in its current form.

18 Chile becomes a full member of the 1967 Treaty of Tlatelolco which bans nuclear weapons from Latin America and the Caribbean.

20– The IAEA says that North Korea has posed too many restrictions on inspectors; the US warns the North to allow inspections of 7 sites (21); North Korean diplomats hold inconclusive talks with the IAEA (24).

26 At UN Conference on Disarmament, talks begin on a Comprehensive Test Ban Treaty.

February

3–   Ukraine's parliament removes conditions it had placed on ratification of START Treaty; US President Clinton promises to double aid to Ukraine to help dismantle nuclear weapons (10).

14   UN General Assembly appoints José Ayala Lasso, from Ecuador, as the first UN High Commissioner for Human Rights.

15–   North Korea agrees to allow inspectors to check 7 acknowledged nuclear sites; US and North Korea reach agreement that the North will admit IAEA inspectors and the US will cancel planned military exercises with South Korea (25).

15   UK and Russia agree to stop targeting strategic nuclear missiles at each other's territory.

18–   UK and China join international ban on dumping nuclear waste at sea; ban comes into force (21).

19   India test fires *Agni*, intermediate-range ballistic missile.

March

1–   IAEA inspectors enter North Korea; South Korea cancels joint military exercises with US (3); IAEA inspectors announce that the North prevented full inspections of key nuclear installations and US cancels high-level talks with North Korea (16); IAEA Board of Governors refers issue to UNSC (21); UNSC calls on North Korea to allow complete IAEA inspections of its nuclear facilities (31).

6   Ukraine sends 60 nuclear warheads to Russia for dismantling.

14   US extends nuclear test ban for another year.

17   US and Russia agree to allow inspections of storage sites of plutonium triggers removed from dismantled warheads and Russia agrees to close 3 plutonium-producing reactors.

25   Signatories of 1989 Convention on Transboundary Movement of Hazardous Wastes (which does not include the US) agree to ban all exports of hazardous waste from industrialised countries to developing ones.

April

4–   North Korea denounces UNSC statement on IAEA inspections; President Kim Il Sung denies existence of nuclear weapons in North Korea (16); tells IAEA it will soon refuel its largest nuclear reactor and invites IAEA to supervise (21); IAEA cancels visit to check refuelling because North Korea refuses to specify IAEA's role (29).

9   Talks between US and Pakistan end without Pakistan's agreement to freeze its nuclear programme.

12   US Congressional report claims Russia has not yet begun to destroy its chemical weapons, as promised in 1990.

15   Representatives of over 120 countries sign final act of Uruguay Round in Marrakesh, Morocco, creating World Trade Organisation (WTO).

May

3   UNSC issues guidelines on the factors to be taken into account before establishing peacekeeping operations.

10   In Tengen, Germany, 5.6g of plutonium-239 are found in the home of a businessman.

12–   Agreement reached between North Korea and the IAEA to allow supervision of removal of fuel rods from nuclear reactor; IAEA inspectors arrive in North Korea (18); IAEA negotiators arrive to discuss storage of fuel rods removed from main reactor (24); they leave after government rejects proposal for storage of spent fuel rods (28); UNSC warns North Korea not to conceal evidence that fuel had been diverted to weapons programme (30).

15   Ukraine pledges to adhere to Missile Technology Control Regime.

20   Indian Prime Minister Rao meets US President Clinton in Washington and both agree need for ban on nuclear testing and production of fissile material.

25   UNSC lifts arms embargo on South Africa.

30   Brazil signs 1967 Tlatelolco Treaty banning nuclear weapons from Latin America and the Caribbean.

June

2–   US confirms it will seek UN sanctions against North Korea over removal of fuel rods from reactor without verification; IAEA suspends technical aid to North Korea (10); the North announces its withdrawal from the IAEA (13); former US President Jimmy Carter begins visit to North Korea, and North Korean President Kim Il Sung offers to suspend nuclear programme to participate in North–South talks if threat of sanctions is dropped and US–North Korean talks resume (15).

4–   India test fires *Prithvi* short-range missile (6).

10   China carries out underground nuclear test

July

8–   North Korean President Kim Il Sung dies; US–North Korean talks postponed; both parties agree to resume talks in August (21).

August

5–   US and North Korea resume high-level negotiations; North Korea agrees to restructure its nuclear reactors and will not reprocess spent uranium fuel rods; both sides say willing to normalise relations (12); South Korea offers aid to restructure North Korea's reactors (15).

9   US decides not to deliver F-16 fighter planes to Pakistan, because of concerns about Pakistan's nuclear programme.

10–   In Munich, Germany, police arrest 3 people who had arrived from Moscow possessing 330g of plutonium-239; German police seize small amount of nuclear material in Bremen (12); Euratom confirms that material seized in Munich was manufactured in Russia (18); Germany and Russia agree to increase cooperation between intelligence agencies to fight nuclear smuggling (22).

23–   Former prime minister of Pakistan, Nawaz Sharif, announces Pakistan possesses atomic bomb; Prime Minister Bhutto denies Sharif's claim (24).

29   Hungarian police arrest two people attempting to sell 2kg of radioactive material, allegedly from the former USSR.

September

5   Third UN Conference on Population and Development opens.

10–   In North Korea, US and North Korea begin low-level talks to discuss establishing liaison offices, as North Korea allows IAEA inspectors to visit 2 previously closed nuclear sites; US and North Korea resume high-level talks firming up outline agreement reached in August (23); defector says North Korea successfully tested 3 nuclear bombs in Russia and Ukraine in 1992 (27); US and North Korea adjourn talks, having made little progress (29).

10–   Russian police seize 100kg of stolen uranium-238; Bulgarian police find 19 containers of radioactive material hidden in two cellars in Sofia (14); 4 Slovaks are caught smuggling 750g of uranium-235 into Hungary (29).

12   UN Secretary-General backs Japan's candidacy for permanent seat on the UNSC.

21   Two-week preparatory conference in Geneva to discuss renegotiation of 1972 convention banning biological weapons opens.

26   Addressing UN General Assembly, Russian President Yeltsin proposes radical cuts in nuclear weapons and materials, while President Clinton proposes limitations on use of anti-personnel mines.

27   At a two-day summit in Washington, the US and Russia agree to seek early ratification of START II Treaty, examine further reductions in remaining nuclear forces, and exchange information on nuclear weapons and fissile materials stockpiles.

October

7–   China conducts underground nuclear test at Lop Nor but says that it will end nuclear tests when negotiations on a CTBT are completed; China and US begin talks on nuclear weapons testing and the NPT (31).

13–   Romanian police seize 7kg of uranium and strontium; in Moscow, police confiscate 27kg of smuggled uranium-238 (19).

21–   US and North Korean negotiators sign accord committing North Korea to abandon its nuclear-weapons programme and to accept IAEA inspections, in return for US diplomatic recognition and aid for updating its nuclear technology and energy alternatives; South Korea announces it will pay over half the aid programme costs (27); US and South Korea announce they will hold smaller joint military exercises instead of major annual military exercise (28).

November

1   UN Trusteeship Council formally suspends operations, since all its territories have achieved independence.

1–   North Korea announces it has halted work on 2 reactors under construction; China promises to help with US-brokered agreement to control North Korea's nuclear programme (2); North Korea announces it has frozen work at its nuclear reactors (18); IAEA representatives begin visit to North Korea (22); IAEA inspectors say the North has halted nuclear programme and stopped building 2 nuclear reactors (28).

3   IAEA convenes meeting of 35 countries to discuss ways to increase secu-

rity of nuclear materials.

16   Ukraine's parliament ratifies NPT.

16   Over 20 countries reach tentative agreement to ban international sales of anti-personnel landmines.

16   UN Convention on the Law of the Sea enters into force.

21–   First UN world conference on transnational organised crime opens in Naples; 136 countries pledge to coordinate fight against syndicates (23).

23   US discloses that it successfully completed airlift of 600kg of weapons-grade uranium from Kazakhstan to the US.

December

5   START I Treaty enters into force, and Ukraine formally accedes to the NPT.

8–   GATT conference confirms that Uruguay Round of GATT and WTO will enter into force on 1 January 1995; US joins with others to reject China's demands to join WTO (20).

8   Head of IAEA, Hans Blix, says North Korea is cooperating fully with IAEA over freezing its nuclear-power industry.

14   In Prague, Czech authorities seize 3kg of highly enriched uranium-235.

GLOSSARY

ABM	Anti-Ballistic Missile
ANC	African National Congress
APEC	Asia-Pacific Economic Cooperation
ARF	ASEAN Regional Forum
ASEAN	Association of South-east Asian Nations
BHC	Bosnia-Herzegovina Command
BiH	Bosnian Government Army
BJP	*Bharatiya Janata* Party (India)
BSA	Bosnian Serb Army
BWC	Biological Weapons Convention
CFSP	Common Foreign and Security Policy
CIS	Commonwealth of Independent States
CJTF	Combined Joint Task Force
COCOM	Coordinating Committee for Multilateral Export Controls
CTBT	Comprehensive Test Ban Treaty
CTR	Cooperative Threat Reduction programme
CWC	Chemical Weapons Convention
ECOWAS	Economic Community of West African States
EFTA	European Free Trade Association
EMU	Economic and Monetary Union
ESDI	European Security and Defence Identity
EU	European Union
EZLN	*Ejército Zapatista de Liberación Nacional* (Mexico)
FIS	*Front Islamique de Salut* (Algeria)
FSU	Former Soviet Union
G-7	Group of Seven
GATT	General Agreement on Tariffs and Trade
GCC	Gulf Cooperation Council
GDP	gross domestic product
GIA	*Groupe Islamique Armée* (Algeria)
GNP	gross national product
GPS	Global Positioning System
HEU	highly enriched uranium
HV	Croatian Army
HVO	Bosnian Croat Army
IAEA	International Atomic Energy Agency

ICFY	International Conference on the Former Yugoslavia
IDF	Israeli Defence Force
IEC	Independent Electoral Commission
IFP	Inkatha Freedom Party
IGC	Inter-Governmental Conference
IRA	Irish Republican Army
LDP	Liberal Democratic Party (Japan)
LEGCO	Hong Kong Legislative Council
LEU	low enriched uranium
LTTE	Liberation Tigers of Tamil Eelam (Sri Lanka)
LWR	light-water reactor
MINATOM	Ministry of Atomic Energy (Russia)
NAFTA	North American Free Trade Agreement
NATO	North Atlantic Treaty Organisation
NIS	newly independent states of the former Soviet Union
NPT	Nuclear Non-Proliferation Treaty
OECD	Organisation for Economic Cooperation and Development
OSCE	Organisation for Security and Cooperation in Europe (formerly CSCE)
PDD	Presidential Decision Directive (US)
PDRY	Peoples' Democratic Republic of Yemen
PFP	Partnership for Peace
PLA	Peoples' Liberation Army (China)
PLO	Palestine Liberation Organisation
PRI	Institutional Revolutionary Party (Mexico)
RDP	Reconstruction and Development Programme (South Africa)
Renamo	*Resistência Nacional Moçambicana*
RPF	Rwanda Patriotic Front
ROK	Republic of Korea (South Korea)
RSK	Republic of Serbian Krajina
SAARC	South Asian Association for Regional Cooperation
SACEUR	Supreme Allied Commander Europe
SANDF	South African National Defence Force
SDF	Self-Defense Forces (Japan)
SDPJ	Social Democratic Party of Japan
SPLA	Sudan Peoples Liberation Army
START	Strategic Arms Reduction Talks

THAAD	Theater High Altitude Area Defense (US)
UN	United Nations
UNHCR	United Nations High Commission for Refugees
UNITA	*União Nacional para a Independência Total de Angola*
UNPA	United Nations Protected Area
UNPROFOR	United Nations Protection Force
UNSC	United Nations Security Council
UNSCOM	United Nations Special Commission on Disarmament (Iraq)
WEU	Western European Union
WTO	World Trade Organisation
YAR	Yemen Arab Republic